Chromatic Monitoring
of Complex Conditions

Series in Sensors

Series Editors: Barry Jones and William B. Spillman

Other recent books in the series:

Series in Sensors

Chromatic Monitoring
of Complex Conditions

Edited by

Gordon Rees Jones
University of Liverpool, UK

Anthony G. Deakin
University of Liverpool, UK

Joseph W. Spencer
University of Liverpool, UK

CRC Press
Taylor & Francis Group
Boca Raton   London   New York

CRC Press is an imprint of the
Taylor & Francis Group, an **informa** business

A TAYLOR & FRANCIS BOOK

CRC Press
Taylor & Francis Group
6000 Broken Sound Parkway NW, Suite 300
Boca Raton, FL 33487-2742

© 2008 by Taylor & Francis Group, LLC
CRC Press is an imprint of Taylor & Francis Group, an Informa business

No claim to original U.S. Government works
Printed in the United States of America on acid-free paper
10 9 8 7 6 5 4 3 2 1

International Standard Book Number-13: 978-1-58488-988-5 (Hardcover)

Library of Congress Cataloging-in-Publication Data

Chromatic monitoring of complex conditions / editors, Gordon Rees Jones,
    Anthony G. Deakin, Joseph W. Spencer.
        p. cm. -- (Sensors series)
    Includes bibliographical references and index.
    ISBN 978-1-58488-988-5 (hardback : alk. paper) 1. Quantum chromodynamics.
    2. Chromatographic analysis 3. Computational complexity. I. Jones, G. R.
    (Gordon Rees), 1938- II. Deakin, Anthony G. III. Spencer, Joseph W. IV. Title.
    V. Series.

QC793.3.Q35C57 2008
681'.25--dc22   2008003951

Visit the Taylor & Francis Web site at
http://www.taylorandfrancis.com

and the CRC Press Web site at
http://www.crcpress.com

Contents

Prologue

Complexity may be considered in the context of a system, as this allows entities and phenomena to be delineated and their interactions to be defined (Checkland, 1984). Natural systems may be viewed analytically in terms of their constituent elements in an infrastructure of biology, chemistry, physics, and mathematics. Additionally, they are being viewed as emergent, synthetic, complex entities (Paton, 1997) with different behavior and functionality emerging at each level in the hierarchy—for example, the cell, the organ, the organism, and the society. Complex natural systems—for example, weather—tend to occur in (or be) dynamical, nonlinear, chaotic environments, and are subject to unpredictability at the edge of chaos (Lewin, 1993; Mitchell Waldrop, 1992), while following certain laws and also attractors.

The complexity found in nature and natural systems is increasingly applied to man-made systems, where technology is designed for the purposes of reliably providing products or services with a high level of automation in a planned and controlled manner—for example, electric power generation, distribution, and transmission systems. Complex man-made systems characteristically have order and stability, but there are possibilities that they may be drawn into a chaotic regime and into failure modes in unforeseen ways. There have been recent examples of such situations with electric power systems in California and Europe (Merlin and Desbrosses, 2007). Furthermore, evaluating the state of the health of complex systems from the instantaneous values of a number of system parameters is also difficult, given that there are, potentially, unpredictable failure modes that may not be taken into account as the combinations of parameter values are too numerous to consider in their entirety, and systems do not operate in isolation.

Generally, whether the system is natural or is man-made and provided with fault tolerance and self-regulation, some level of direct human monitoring and, ultimately, intervention in the system, is required. In the case of man-made systems, this ranges in immediacy and frequency from ad hoc to regular inspection of the system and system data, and from scheduled maintenance to manually overriding controls.

Technology drive has a further implication with respect to the requirements for monitoring. It fuels opposing tendencies that need to be reconciled—on the one hand, the tendency towards greater system intelligence and autonomy (freeing humans from supervision and intervention at some levels) and, on the other hand, towards increasing complexity and potential unpredictability (requiring increased expert human supervision, especially for rare and unusual, but major, events). Arguably, the gap between these tendencies can be bridged through intelligent monitoring systems that can be applied to the whole monitored complex system holistically to assess and assure overall system behavior. Desirable features of intelligent monitoring systems are that they address the whole system, with information fusion from an optimum amount of sensor data; are proactive and prognostic; are interactive in supporting the human expert/operator and can be queried when necessary; are nonintrusive, and thereby are easily retrofitted; and are generically applicable, including the capability for using naturally available data about the system, e.g., from ambient acoustic, vibration, temperature, color, and radio-frequency information, as well as data from sensors specifically introduced, or even from existing dial readings, in order to maximize cross-correlation of information on system status.

When considering the whole system, a traditional array of sensors may not be sufficient to ascertain overall system health and behavior, representing instead a selection of possible symptoms. A holistic approach may make use of naturally occurring information such as patterns in the acoustic, temperature, light, infrared, electromagnetic, radio-frequency, etc., domains to complement an existing sensor array, or may even derive additional information from existing sensor data. For example, the center of gravity of an aircraft in flight, which cannot be measured directly, may be deduced from other sensor data used as indirect indicators. In health care, temperature (normally self-regulating through homeostatic mechanisms) can be an important indicator of health. However, a medical practitioner takes into account other, predominantly visual (and also verbal) information before considering temperature, aiming to perform an overall health check. The aim is to check and remediate any existing problems (diagnosis) and, in so doing, to anticipate and prevent further problems (prognosis) before they require emergency intervention.

This book describes and explains one approach, the chromatic methodology (Jones et al., 2000; Jones et al., 2003), for the intelligent monitoring of complex systems. It is demonstrated that chromatic processing is analogous to human vision while extending far beyond the visible part of the spectrum and to other domains such as the acoustic. The approach and its capabilities are described along with examples of graphical polar plots, which may depict, for example, process life cycles, some having "centers of gravity." The examples are drawn from diverse areas such as nonintrusive behavior monitoring; the balance of gases during an anaerobic digestion cycle; high voltage transformer gases; balances of pollution from various sources; information contained in acoustical signals from mechanical systems such as machine bearings, vehicle wheels, etc.; the control of electrical plasmas; probing biological tissues noninvasively; and diagnosing liquids used in industrial plants. There is also consideration of how the properties of thin films of materials and materials that polarize light can be used for chromatic sensing of various physical and chemical parameters.

References

Checkland, P. B. (1984). *Systems Thinking, Systems Practice*, John Wiley, Chichester, U.K.

Jones, G. R., Deakin, A. G., Yan, J., and Spencer, J. W. (2003). *Prognostic Intelligent Monitoring of Energy Systems*, in Ed. Colette R. Benson, *Proc. Euro TechCon 2003*, pp. 157–178, November 5–6, 2003, Manchester, TJ/H2b Analytical Services, Inc.

Jones, G. R., Russell, P. C., Vourdas, A., Cosgrave, J., Stergioulas, L., and Haber, R. (2000). The Gabor transform basis of chromatic monitoring, *Meas. Sci. Technol.*, 11, 489–498.

Lewin, R. (1993). *Complexity: Life at the Edge of Chaos*, J. M. Dent, London.

Merlin, A. and Desbrosses, J. P. (2007). European incident of 4th November 2006: The events and the first lessons drawn, *Electra*, 230, 4–11.

Mitchell Waldrop, M. (1992). *Complexity: The Emerging Science at the Edge of Order and Chaos*, Viking, London.

Paton, R. C. (1997). Glue, verb and text metaphors in biology, *Acta Biotheoretica*, 45, 1–15.

Acknowledgments

There are several organizations which, over several years, have interacted to enable the chromatic methodology concepts to be deployed and tested in a variety of monitoring applications, many of which are presented in this book. Their contributions are much appreciated.

AMEC plc
ARRIVA plc
Aspects and Milestones, Dementia Care Home, Bristol, U.K.
BICC
BNFL
British Council
Bulgarian Academy of Sciences
Chell Instruments
Dieline
EPSRC (U.K.)
European Union Regional Development Fund
Health and Safety Executive (U.K.)
Global Technology Research
London Borough of Merton
Liverpool City Council
Liverpool Womens Hospital
Lucas
Malvern Instruments
Magnetic Resonance and Image Analysis Research Centre, Liverpool, U.K.
NEI Reyrolle
NGC
Pernix
Royal Society (U.K.)
Sefton Borough Council
Shell Research
Tekgenuity
University of Liverpool

The provision of data by the late Dr. Graham Nicholson (Institute of Orthopaedics and Musculoskeletal Science, UCL) (Chapter 8) and C. Yan (University of Liverpool; Chapter 11) is acknowledged.

The inputs of Dr. I. Moumdjiev (Clinic of Paediatrics) and Dr. E. Milieva (Department of Theoretical Physics), both of the University of Plovdiv (Bulgaria), are acknowledged in relation to Chapter 8; Dr Liu (Centre for Wavelets, Approximation and Information Processing) of the University of Singapore is acknowledged in relation to Figure 10.4.3 (Chapter 10).

The technical input of Dr. Duncan Smith and Nicola Telfer is much appreciated.

The administrative and secretarial assistance provided by S. Kallio and L. Johnson in the preparation of the manuscript is acknowledged.

List of Contributors

E. Borisova
Centre of Biomedical Engineering
  Prof. Ivan Daskalov
Bulgarian Academy of Sciences
Sofia, Bulgaria

A. G. Deakin
Centre for Intelligent Monitoring
  Systems
Dept. of Electrical Engineering and
  Electronics
University of Liverpool
Liverpool, UK

B. Djakov
Institute of Electronics
Bulgarian Academy of Sciences
Sofia, Bulgaria

C. A. Egan
Acoustics Research Unit
School of Architecture
University of Liverpool
Liverpool, UK

G. R. Jones
Centre for intelligent Monitoring
  Systems
Dept. of Electrical Engineering and
  Electronics
University of Liverpool
Liverpool, UK

A. Koh
Centre for Intelligent Monitoring
  Systems
Dept. of Electrical Engineering and
  Electronics
University of Liverpool
Liverpool, UK

Y. R. Kolupula
Centre for Intelligent Monitoring
  Systems
Dept. of Electrical Engineering and
  Electronics
University of Liverpool
Liverpool, UK

H. M. Looe
Centre for Intelligent Monitoring
  Systems
Dept. of Electrical Engineering and
  Electronics
University of Liverpool
Liverpool, UK

P. Pavlova
Institute of Electronics
Bulgarian Academy of Sciences
Sofia, Bulgaria

C. D. Russell
Consultant
Ex Medical Lasers Unit
University of Liverpool
Liverpool, UK

J. W. Spencer
Centre for Intelligent Monitoring
  Systems
Dept. of Electrical Engineering and
  Electronics
University of Liverpool
Liverpool, UK

K. J. Wong
MAST Group LTD.
Bootle
Merseyside, UK

S. Xu
Information Engineering College
Guangdong University of
    Technology
Guangzhou
Guangdong, P. R. China.

Y. Yokomizu
Department of Electrical
    Engineering and Computer
    Science
Nagoya University
Nagoya, Japan

J. Zhang
Department of Computer Science
Loughborough University
Loughborough
Leicestershire, UK

X. Zhang
Research Institute of Diagnosis
    and Cybernetics,
School of Mechanical Engineering,
Xi'an Jiaotong University
Xian, P. R. China

Section A

Basic Principles

1

Basic Principles of Chromatic Monitoring

G.R. Jones and J. Zhang

CONTENTS

1.1   Introduction

To appreciate the problems of monitoring complex conditions it is useful to consider the concepts of measurement, diagnosis, and monitoring. Measurement is concerned with the quantification of extent by comparison with a standard. As such, it is reductionalist in nature being based upon addressing isolated and specific parameters to a high degree of accuracy.

Diagnosis is concerned with the determination of a condition from a symptom and so implies the establishment of information.

Monitoring is concerned with observation to indicate the likelihood of a condition evolving. Thus, it offers the possibility of addressing the emergent nature of complexity, which involves the interaction of several events, often with the outcomes not anticipated.

Consequently, diagnosis and monitoring may be regarded as bridging the gap between measurement and the production of relevant information that emerges from the complexity of various situations.

A simple example of a problem in monitoring emergence is in the processing of semiconductor materials with electrical plasma (Chapter 4). The process is often monitored via changes in the relative light intensities at each of two specific wavelengths. However, the process is highly susceptible to impurities that produce changes distributed over relatively extensive wavelength ranges away from those being monitored. There are subtle distributed changes that only gradually emerge from the "noise" background and can be missed by conventional reductionalist approaches.

The need to represent information about a complex condition in a manner convenient for assimilation is also important.

With reductionalist data, because of its narrow and specific focus, simple graphical representations are sufficient, e.g., linearizing the relationship between two completely decoupled parameters. With emergent information there is a need to blend simplicity of presentation with a range of different but relevant information. An example is the manner in which a substantial amount of different types of information is embedded in cartography. Subtle coloration is used to distinguish between subsea depths and land altitude above sea level, whereas carefully chosen symbols are used to distinguish between different features such as roads, rail networks, or regional boundaries.

It is within this context of appreciating the need to encompass both reductionalist and emergent methods to deal with complex situations that chromatic monitoring has a role. It respects the need for flexibility in responding to the less expected, for providing quantification for judgmental purposes, and for conveying information, rather than simply data, in a manner suitable for assimilation.

The chromatic methodology is based upon comparisons, which are translated into mathematical cross correlations. Because it is quantitative in nature, it enables numerical scales of judgment to be established. However, unlike methodologies such as neural networks, it has a high degree of traceability, lends itself to a hierarchical approach to different levels of detailed information, and significantly enables the quantified information to be easily assimilated via patterns so that in-depth mathematical knowledge is not mandatory.

1.2 The Nonorthogonal Nature of Chromaticity

Based upon scientific rigor to analyze and segregate, each measurement (e.g., pressure, temperature) in metrology is addressed separately to provide high accuracy for narrowly specific data. Similarly, in telecommunications, cross talk between different communication channels is avoided to prevent data corruption. These two examples which are typical of conventional approaches are orthogonal in that they segregate to achieve exclusivity for predefined situations.

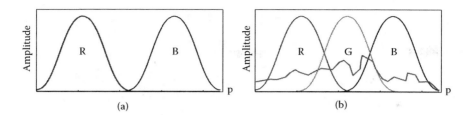

FIGURE 1.1
Responses of receptors (R, G, B) as a function of parameter p: (a) orthogonal receptors; (b) nonorthogonal receptors and data stream.

Such systems may be depicted as shown on Figure 1.1a, whereby the responses of two signal receptors (R, B) cover two separated parts of a parameter range, p (where p, for example, may be frequency). As such the receptors respond to different events independently. The responsivities of the two receptors of Figure 1.1a are said to be orthogonal.

Chromatic systems accommodate a controlled amount of interactivity between channels and are nonorthogonal in nature. Human perception relies upon such interactivity both in art (e.g., polychromatic colors) and in music (e.g., chromatic scales). Although not previously recognized as such, other human sensations of smell, taste, and touch also appear to encompass chromaticity.

More recently, scientific analysis and perspectives of chromaticity have evolved ranging from the concept of the photic field described by Moon and Spencer (1981), the formation of science concerned with color technology (Jones and Russell, 1993), interpretation in terms of Gabor transforms (Stergioulas et al., 2000; Jones et al., 2000) and general monitoring applications (Brazier et al., 2001).

The orthogonal system of Figure 1.1a may be converted into a nonorthogonal system through the introduction of a third receptor, G, whose response bridges the gap between the first two receptors and overlaps their responses as shown in Figure 1.1b. The characteristic shown in Figure 1.1b is typical of an R, G, B tristimulus chromatic system.

The tristimulus system has some important properties for extracting information from complex signals (Jones et al., 2000), which can be appreciated by considering the superposition of the responses of the three R, G, B receptors upon an irregular varying data stream as shown in Figure 1.1b. The outputs of R, G, B (which can be defined mathematically as shown in appendix 1.1) in response to the irregular data stream have overlapping components that contain significant levels of information. This may be illustrated via the colored images shown in Figure 1.2 that are records of a scene obtained via three different optical filters: cyan (Figure 1.2a), yellow (Figure 1.2b), and magenta (Figure 1.2c). When each image is viewed separately, fragments of information are apparent. However, when the three images are superimposed, considerable information emerges (Figure 1.2d), which is greater than

(a)   (b)

(c)   (d)

FIGURE 1.2
(See color insert following page 18). Illustration of the information gain (d) by the superposition of the outputs from three non-orthogonal processors (a, b, c).

that discerned from the sequential viewing of the individual images. This illustrates the power of superposition in enhancing information without loss of breadth. The example relates to how color print images are produced.

1.3   Chromatic Transformations

To extract meaningful information from the outputs of such nonorthogonal data processors it is advantageous to transform these outputs (i.e., R, G, B) into mathematical forms that can emphasize and distinguish particular information being sought. For example, there may be a need to distinguish between a group of signals whose members are all different but each has a variation with a frequency of the same bell-shaped distribution shown in Figure 1.3a. This is the well-known normal or Gaussian distribution frequently encountered in statistics. The Gaussian distribution is defined mathematically in terms of only three parameters, which are its amplitude (P_o), the location of its maximum value (fm), and its width (∇f) according to Equation 1.1 (e.g., Papoulis and Pillai, 2002):

$$P = P_o \exp[-\{(f - fm)/(\sqrt{2}\,\nabla f)\}^2]$$

(1.1)

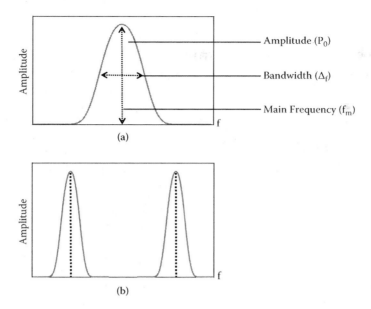

FIGURE 1.3
Gaussian signals: (a) single Gaussian signal; (b) two Gaussian signals.

Thus, Gaussian signals governed by Equation 1.1 may differ from each other in the magnitudes of P_o, fm, ∇f. Consequently, to distinguish between such signals, the outputs of three nonorthogonal processors, R, G, B, would need to be transformed into a form that could yield values of P_o, fm, ∇f.

There has been a considerable amount of work undertaken by color scientists in effectively addressing such issues with respect to color discrimination by the receptors of the human eye (e.g., Levkowitz and Herman, 1993; Rogers, 1985). It is, therefore, advantageous to build upon such a wealth of knowledge in developing the concepts of chromaticity of which color discrimination may be regarded as a simple degenerate form (Jones et al., 2000).

There are many basic transformations well established in color science and that can be conveniently adapted for chromatic monitoring and processing. Of these the H, L, S; H, S, V; L, a, b; and x, y, z methods (e.g., Levkowitz and Herman, 1993; Rogers, 1985; Billmeyer and Saltzman, 1981) have been proven to be particularly useful. The H, L, S transformation is described in Section 1.3.1 and Chapter 2. The H, S, V; x, y, z; and L, a, b transforms are described in Chapter 3.

1.3.1 Basic *H, L, S* Transformation

The outputs of the three nonorthogonal processors (R, G, B) (Appendix 1.I) may be transformed into the chromatic parameters H, L, S according to the following relationships (e.g., Xu and Jones, 2006):

$$H = 240 - 120.g/(g + b) \quad r = 0$$
$$= 360 - 120.b/(b + r) \quad g = 0 \tag{1.2}$$
$$= 120 - 120.r/(r + g) \quad b = 0$$

$$L = (R + G + B)/3 \tag{1.3}$$

$$S = [\max (R, G, B) - \min (R, G, B)]/[\max (R, G, B) + \min (R, G, B)] \tag{1.4}$$

where
$$r = R - \min (R, G, B) \tag{1.5}$$
$$g = G - \min (R, G, B) \tag{1.6}$$
$$b = B - \min (R, G, B) \tag{1.7}$$

Max (R, G, B) and min (R, G, B) represent the parameter (R, G, B) having the highest and lowest values, respectively.

With regard to a Gaussian signal (Equation 1.1, Figure 1.3a), L is an effective signal strength (αP_o, ∇f), H is the dominant value of the parameter (f) and $(1 - S)$ is the equivalent width (∇f). In color science it is the convention to form a polar diagram whereby the azimuthal angle represents H (0–360°), the radius S (0–1), and the vertical axis L (Levkowitz and Herman, 1993), (Figure 1.4a). The case of $S = 1$ corresponds to an infinitely narrow (monochromatic) signal,

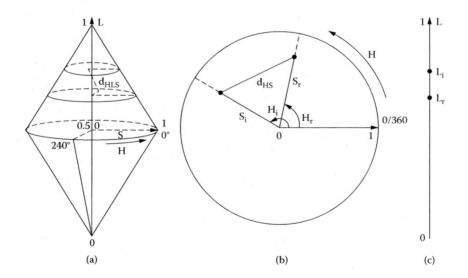

FIGURE 1.4
Changes in chromatic parameters defined on *H–S* and *L* diagrams: (a) *H-L-S*; (b) *H-S*; (c) *L*. (From Xu, Z. and Jones, G. R. (2006), *Meas. Sci. Technol.*, 17, 3204–3211. With permission.)

whereas $S = 0$ implies an infinitely wide signal (i.e., equal amplitudes at all frequencies without a dominating frequency). The case of $H = 0°$ corresponds to a low frequency signal and $H = 270°$ a high frequency one.

If the signal processed by (R, G, B) is not a simple Gaussian signal (Figure 1.3a) but, for example, is composed of two separated Gaussian signals (Figure 1.3b), the values of H, L, S may be regarded as representing an equivalent single Gaussian signal. If the separation of the two Gaussian signals is sufficient in the parameter domain (f), the dominant H value lies in the range $270° < H < 360°$. Consequently, the signal in this range is not of a single Gaussian form (Jones et al., 2000).

This argument may be extrapolated to any signal form (e.g., Figure 1.1b) with the interpretation that the values derived for H, L, S represent another Gaussian of a group of signals of which that signal is a member. In other words, the H, L, S transformation may be regarded as assigning a signal to a particular Gaussian group compatible with the form of the three nonorthogonal processors.

This also provides an insight into the signal discriminating capabilities of the approach. Stergioulas et al. (2000) have shown a capability of discriminating between 95% different signals, which is compatible with the substantial color discrimination capabilities of the human color vision ($\geq 10^6$). Greater discrimination is achievable through the use of additional processors but with little gain obtained with more than six, which are estimated to be capable of discriminating 99.5% of the signals (Stergioulas et al., 2000). Consideration of such extrapolations is made in Section 2.3 (Chapter 2).

1.3.2   *H, L, S* Adaptations

In deploying the H, L, S transformation for chromatic monitoring, there are various adapted forms that have proved useful.

(a)   *Vector Chromaticity (Polar Form)*

Overall chromatic changes defined in terms of H, L, S values can be discriminated, as in color science, in terms of Euclidean distance (Wesolkowski and Jernigan, 1999; Dony and Wesolkowski , 1999), vector angle (Wesolkowski and Jernigan, 1999; Dony and Wesolkowski , 1999), or their combinations (Wesolkowski and Jernigan, 1999; Carron and Lambert, 1994). However, Euclidean distance is not very sensitive to variations in H and S. Vector angle has smaller variations in L, and when the RGB values are low or S is approximately zero, the difference in H becomes irrelevant. Some combination schemes do not have identical scales for H and L differences.

An alternative approach that has proven useful for chromatic monitoring is to divide the H, L, S space (Figure 1.4a) into a 2-D H-S polar plot (Figure 1.4b) and a 1-D L plot (Figure 1.4c). Chromaticity changes d_{HLS} can be quantified via the H-S and L plots individually, i.e., via the component chromaticity changes

d_{HS}, d_L, respectively. These are defined as follows (Xu and Jones, 2006):

$$d_{HS} = \left[(S_i \cos H_i - S_r \cos H_r)^2 + (S_i \sin H_i - S_r \sin H_r)^2 \right]^{1/2} \qquad (1.8)$$

$$d_L = |L_i - L_r| \qquad (1.9)$$

where H_r, L_r, and S_r are the initial values of the chromatic parameters, H_i, L_i, and S_i are the changed values. S and L are orthogonal in 3-D chromaticity space so that the H-S and L plots are also orthogonal. Thus, the total chromatic difference between two signals d_{HLS} is given by the geometrically combined chromaticity differences on the H-S and L diagrams, i.e.,

$$d_{HLS} = \left(d_{HS}^2 + d_L^2 \right)^{1/2} \qquad (1.10)$$

In practice, the numerical value of H is unreliable if S is low (i.e., the signal is approximately uniformly distributed across the spectral range). In such cases, H may be assigned an unreal value, e.g., –199 (Xu and Jones, 2006). Thus, when performing a comparison between two chromatic signatures, if either of their H values is –199, their H values are considered to be identical, so that d_{HLS} is given by

$$d_{HLS} = \left[(L_i - L_r)^2 + S_r^2 \right]^{1/2} \qquad (1.11)$$

If, in addition, $S_r = S_i = 0$, then

$$d_{HLS} = d_L = |L_i - L_r| \qquad (1.12)$$

when both H values are –199 (i.e., both S values are zero).

(b)   Vector Chromaticity (Cartesian Form)

From a chromatic monitoring aspect, the H, L, S parameters derived with Equation 1.2 through Equation 1.7 may be regarded in terms of Cartesian rather than polar coordinates. The implication is that H as well as L and S become rectilinear (Figure 1.5) with a scale 0–360°. As the concern is not with replicating true colors but quantifying chromatic change, another chromatic vector may be defined with such a Cartesian chromatic space according to Equation 1.13 (Rallis et al., 2005):

$$C_T = \sqrt{a_H H_N^2 + a_L (b_L - L_N)^2 + a_S (b_S - S_N)^2} \qquad (1.13)$$

where a_H, a_L, a_S are weighting factors for H, L, S, respectively; H_N, L_N, S_N are normalized values of H, L, S, i.e.,

$$H_N = H/H_0, \quad L_N = L/L_0, \quad S_N = S/S_0 \qquad (1.14)$$

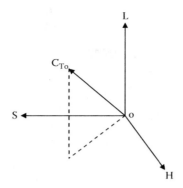

FIGURE 1.5
Cartesian coordinates H, L, S chromatic diagram. ($C_{T_0} = C_T$ with $b_L = b_S = 0$, $a_H = a_S = a_L = 1$ equation 1.13)

whereby H_0, L_0, S_0 are reference values $> H, L, S$, respectively, and

$$b_L = 0 \quad \text{if } dH/dL \text{ positive}$$
$$ 1 \quad \text{if } dH/dL \text{ negative} \tag{1.15}$$

$$b_S = 0 \quad \text{if } dH/dS \text{ positive}$$
$$ 1 \quad \text{if } dH/dS \text{ negative} \tag{1.16}$$

The values of a_H, a_L, a_S are determined according to a number of factors including the signal-to-noise ratio of H, L, S, different susceptibility of each chromatic parameter H, L, S to other factors, etc.

As an example of considerations in applying the above equations, if the monitoring is particularly susceptible to random lightness variations, then a_L may be set to zero, leaving C_T dependent on only H, S, because, according to the chromaticity parameter definitions (e.g., Xu and Jones, 2006), H, S are independent of lightness L.

(c)   *2-D Polar Diagrams*

It can often be preferable to display H, L, S as two, 2-D polar diagrams (Figure 1.6; Jones et al., 2005) rather than as a 3-D polar diagram (Figure 1.4a) to simplify the observation of trend patterns. For example, visualizing the variation of H with L and S separately under time-varying conditions can identify different operational domains.

(d)   *Second-Generation Chromaticity*

For many chromatic monitoring applications the chromatic parameters H, L, S may reflect the variation of a condition with time, space, etc. Under such conditions, second-generation chromatic processing (Zhang et al., 2005;

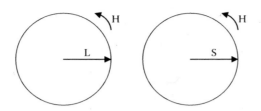

FIGURE 1.6
2-D *H-L*, *H-S* polar diagrams.

TABLE 1.1

Summary of Various *H*, *L*, *S*-Based Chromatic Schemes

H, L, S Form	Figure	Equation	Comment
Basic	1.4a	1.2–1.7	3-D *H-L-S* polar
Vector polar	1.14b,c	1.10	Resultant vector d_{HLS}
Vector Cartesian	1.5	1.13	Resultant vector (C_T)
2-D polar	1.6	1.2–1.7	*H-L*, *H-S* polar
Second generation	1.6	1.2–1.7	Repeated *H, L, S* transform

Zhang and Jones, 2005) may be implemented, whereby each of the primary *H*, *L*, *S* parameter's variation with time may be processed again by applying nonorthogonal processors to each in the time, space, etc., domains. This produces an additional three chromatic parameters for each of the primary *H*, *L*, *S* parameters but not all of which may contain relevant information. The process can, however, be useful for compressing time-based data to provide relatively simple signatures of the behavior of a system.

It is also possible to apply such second-generation processing to variations of *H* with *L* and *S*. This enables patterns of chromatic variations to be quantified in terms of coordinates represented by second-generation values of *H, L, S*.

The various *H*, *L*, *S*-based schemes for chromatic processing are summarized on Table 1.1.

1.4 Summary

Some basic concepts of the chromatic methodology have been described.

The method is related to the photic field description of Moon and Spencer (1981) and the Gabor transforms of signals (Jones et al., 2000).

It is based upon the cross correlation of a number of data channels.

The approach can be used for extracting information about complex conditions that may be evolutionary in nature.

The *H*, *L*, *S* algorithms of color science provide the basis of a convenient means for processing the outputs from chromatic processors.

With such algorithms, it can be demonstrated how the human perception of color is a special, degenerate case of chromaticity.

The chromatic principles can be applied to monitoring domains other than optical wavelength such as time, space, and vibration.

Appendix 1.I   Output of a Chromatic Processor

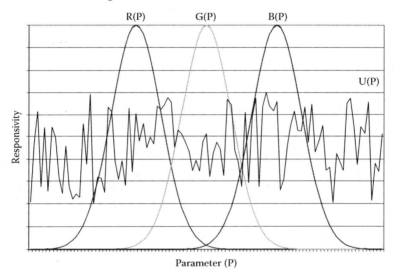

FIGURE APP 1.1

Detectors responses $R(P)$, $G(P)$, $B(P)$ are a function of the parameter (P). $U(P)$ is the amplitude of a signal that varies with the parameter (P).

The output of a detector $X(P)$ addressing the signal $U(P)$ is

$$X_o = \int_P X(P) \cdot U(P) dP \qquad (1.I.1)$$

There are a number of special cases:

1. Monochromatic signal

$$U(P) = U(P_1) \text{ at } P_1 = 0 \text{ otherwise}$$

$$\therefore X_o = X(P_1) \cdot U(P_1) \qquad (1.I.2)$$

2. Equal amplitude signal

$$U(P) = U_o$$

$$\therefore X_o = U_0 \int_P X(P) dP \qquad (1.I.3)$$

3. Discrete signals

$$U(P) = \sum_P U(P)$$

$$\therefore X_o = \sum_P X(P) \cdot U(P) \tag{1.I.4}$$

4. Triangular processor with equal amplitude signal

$$X(P) = mP + c$$

$$\therefore X_o = U_o[(mP^2/2) + cP]_{P_1}^{P_2} \tag{1.I.5}$$

References

Brazier, K.J., Deakin, A.G., Cooke, R.D., Russell, P.C., and Jones, G.R. (2001). Colour space mining for industrial monitoring, in *Data Mining for Design and Manufacturing*, Braha, D., Ed., *Massive Computing Series*, Vol. 3, chap. 16.

Billmeyer, F.W. and Saltzman, M. (1981). "Principles of Color Technology" John Wiley, New York.

Carron, T. and Lambert, P. (1994). Colour edge detector using jointly hue, saturation and intensity, *IEEE Int. Conf. on Image Process*, pp. 977–981.

Dony, R.D. and Wesolkowski, S. (1999). Edge detection on colour images using RGB vector angle, *Proc. IEEE CCECE '99* (Canada). Edmonton, Canada, May 9–12, 1999.

Jones, G.R. and Russell P.C. (1993). Chromatic modulation based metrology, *Pure Appl. Opt.* 2, 87–110.

Jones, G.R., Deakin, A.G., and Spencer, J.W. (2005). Multistimulus chromatic processing of complex signals, *Proc. Complex Systems Monitoring Session, Int. Complexity, Sci. Society Conf.* (Liverpool), pp. 5–15.

Jones, G.R., Russell, P.C., Vourdas, A., Cosgrave, J., Stergioulas, L., and Haber, R. (2000). The Gabor transformation basis of chromatic monitoring, *Meas. Sci Technol.*, 11, 489–498.

Levkowitz, H. and Herman, G.T. (1993). GHLS: A general lightness, hue and saturation colour model CVGIP, *Graphic Image Processing* 55, 271–285.

Moon, P. and Spencer, D.E. (1981). *The Photic Field*, MIT Press, Cambridge, MA.

Papoulis, A. and Pillai, S.H. (2002). *Probability, Random Viariables and Stochastic Processes*, 4th ed., McGraw-Hill, New York.

Rallis, I., Gloman, L., Deakin, A.G., Spencer, J.W., and Jones, G.R. (2005). Polychromatic monitoring of complex biological systems. *Proc. Complex Systems Monitoring Session, Int. Complexity, Sci. Society Conf.* (Liverpool) p. 40–45.

Rogers, D. (1985). *Procedural Elements for Computer Graphics*, McGraw-Hill, New York.

Stergioulas, L., Vourdas A., and Jones, G.R. (2000). Gabor representation of a signal using a truncated Von Neumann lattice and its practical implementation using chromatic measurement, *Opt. Eng.*, 39, 1965–1971.

Wesolkowski, S. and Jernigan, E. (1999). Colour edge detection in RGB using jointly Euclidean distance and vector angle, *Proc. IAPR Vision Interface* (Canada), pp. 9–16. Trois–Rivieres, Canada, May 1999.

Xu, Z. and Jones, G.R. (2006). Event and movement monitoring using chromatic methodologies, *Meas. Sci. Technol.,* 17, 3204–3211.

Zhang, J. and Jones, G.R. (2005). Chromatic processing for event probability description, *Proc. Complex Systems Monitoring Session, Int. Complexity, Sci. Society Conf.* (Liverpool), pp. 61–67.

Zhang, J., Jones, G.R., Deakin, A.G., and Spencer, J.W. (2005). Chromatic processing of DGA data produced by partial discharges for the prognosis if HV transformer behaviour, *Meas. Sci. Technol.* 16(2), 556–561.

2

Characteristics of Chromatic H,L,S Systems

A.G. Deakin, Y.R. Kolupula, A. Koh, H.M. Looe, and J. Zhang

CONTENTS

2.1   Introduction

The basic concepts of chromaticity may be used for establishing operational characteristics for various monitoring applications. Such aspects may be considered in two respects:

(a) The manner in which chromatic processor characteristics may be changed and the effect of this on monitoring

(b) Treatment and use of discrete data sets

Processor characteristics considerations include the effects of the processor profile, extent of nonorthogonality, etc., on the output chromatic coordinates (H, L, S). It concerns the use of multiple processors in addition to the three (tristimulus) conventionally used. It addresses the question of optimum and maximum numbers for practical deployment. Adaptation of processors for 2- and 3-D monitoring in the space domain is also considered.

Discrete data sets enable data from different sensors or sources to be manipulated. These are separate data elements in the parameter domain rather than a continuous signal. This impacts upon the processor characteristics that can be advantageously deployed. Chromatic analysis of such discrete data sets enables the probability of particular events or conditions to be addressed. It is also possible to explore different kinds of information that might be contained within a data set representing complex conditions and identified on chromatic polar diagrams.

2.2   Effect of Different Processor Response Profiles on H, L, S Chromatic Space

2.2.1   Introduction

In its basic form, chromatic processing involves addressing a signal with three nonorthogonal (overlapping) processors (tristimulus system; Chapter 1). The information that appears following such a procedure depends upon the characteristics of the processors, e.g., their degree of nonorthogonality, their profile form, width, and relative amplitudes.

These effects may be perceived in visual images recorded via different optical filters with particular characteristics to modify the spectral responses of the detectors. Examples of the resultant images are given in Figure 2.2.1. Figure 2.2.1a shows a basic image with no intervening filter, whereas Figure 2.2.1b and 2.1c show the same image recorded via green filters with different wavelength-dependent profiles. The examples show how changing a processor's response profile in one aspect affects the color balance and so the information perceived in the image.

(a)

(b)

(c)

(d)

COLOR FIGURE 1.2
Illustration of the information gain (d) by the superposition of the outputs from three non-orthogonal processors (*a, b, c*).

(a)

(b)

(c)

COLOR FIGURE 2.2.1
Effect of changes in chromatic processors response profiles on image perception: (a) Basic image; (b) G processor modified; (c) G processor modified further.

(a)   (b)

COLOR FIGURE 2.5.1
Color image of tree foliage: (a) spring/summer; (b) autumn. Courtesy Trevor Morris, Trevor Morris photographics.

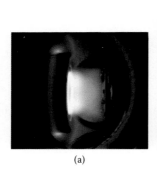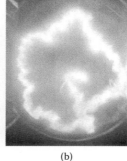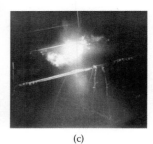

(a)   (b)   (c)

COLOR FIGURE 4.1.1
Examples of technological plasmas: (a) Plasma used for processing materials (Courtesy Professor J. Bradley); (b) convoluted electric arc plasma in a rotary arc circuit breaker; (c) electric arc plasma on a high-voltage railway power line. (From Jones, G.R., Spencer, J.W., and Yan, J., in *Advances in High Voltage Engineering*, IEE, London, 2004, pp. 545–590. With permission.)

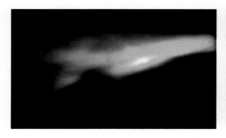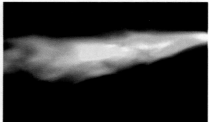

COLOR FIGURE 4.2.6
Sequential images of a plasma jet with particles (air—75 slm, 200 A; exposure—0.5 ms; interframe time—several ms).

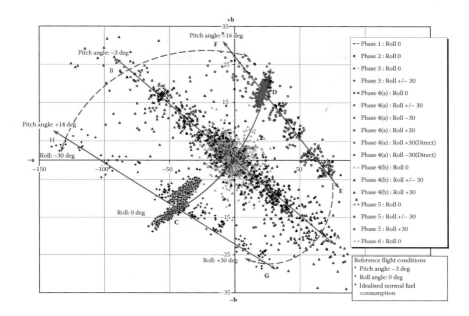

COLOR FIGURE 5.4.3
CIE Lab chromatic diagram of aircraft attitude variation during various phases of flight simulation. (From Anupriya, 2001, *Multi-Sensor Data Fusion for Aircraft Fuel Systems Using Chromatic Processing*, Ph.D. thesis, University of Liverpool.)

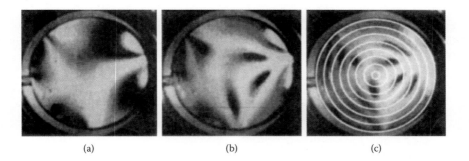

(a)   (b)   (c)

COLOR FIGURE 6.1.1
Images of a photoelastic diaphragm with stresses produced by various pressure differences across the diaphragm: (a) 20 lb/in.^{2}; (b) 40 lb/in.^{2}; (c) eight annular rings used for 2-D analysis (Ahmed, S. U., Intelligent Remote Chromatic Processing, Ph.D. thesis, University of Liverpool, U.K., 1998.)

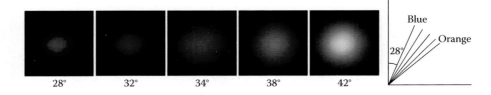

COLOR FIGURE 6.3.4.1
Chromatic images as a function of polarization angle (28–42°) for maltodextrine (Egan, 2006). (Sketch shows relative inclinations of the polarization for each image).

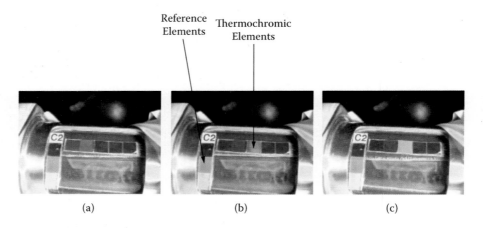

COLOR FIGURE 6.3.4.4
Images of thermochromic elements at different temperatures. (Temperature increasing in the order (a), (b), (c) as chromaticity of elements changes in the order 2, 3 from the left.)

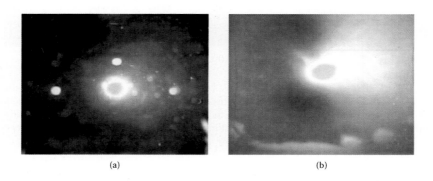

COLOR FIGURE 7.2
Images of forward-scattered light from a water medium. (a) No particulates, (b) 10-μm particulates.

(1)   (2)   (3)

COLOR FIGURE 7.4
Images of forward-scattered light from particles formed by electric arcs in a SF6 high-voltage
circuit breaker.

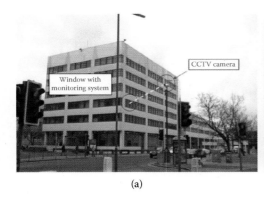 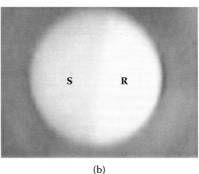

(a)   (b)

COLOR FIGURE 9.2.2
Remote CCTV camera monitoring of a particulates accumulator unit. (a) View of system, (b)
typical zoomed view of particle filter (R—Reference, S—Signal). (From Jones, G. R., Spencer,
J. W., Reichelt, T. E., Aceves-Fernandez, M., and Kolupula, Y. R. [2005]. Air Quality Monitor-
ing in the City of Liverpool Using Chromatic Modulation Techniques. CATCH Final Project
Report, University of Liverpool. With permission.)

COLOR FIGURE 9.2.4
Images of 10-μm particulates-loaded filter from a city center bus terminus showing changes
in chromatic H. (Lamp voltage –3.5 V, filter—micro fiber 50). (From Jones, G. R., Spencer, J. W.,
Reichelt, T. E., Aceves-Fernandez, M., and Kolupula, Y. R. [2005]. Air Quality Monitoring in
the City of Liverpool Using Chromatic Modulation Techniques. CATCH Final Project Report,
University of Liverpool. With permission.)

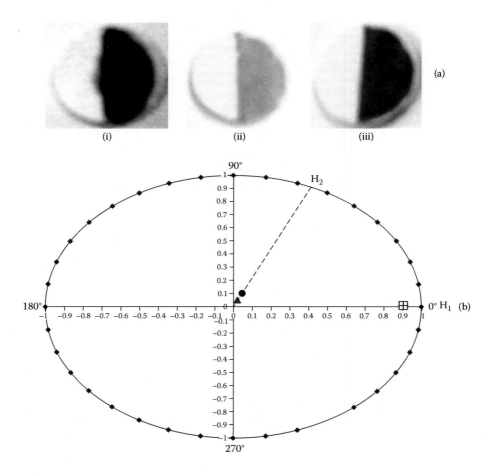

COLOR FIGURE 9.2.5
Chromaticity of different particle types accumulated on micropore filters. (a) Images of signal and reference parts of filters—(i) urban air, (ii) incense smoke, (iii) tobacco smoke; (b) H–S polar diagram for different particles. ▲—urban air, ⊞—tobacco smoke, ●—incense smoke. (From Kolupula, Y. R [2007]. Private Communication.)

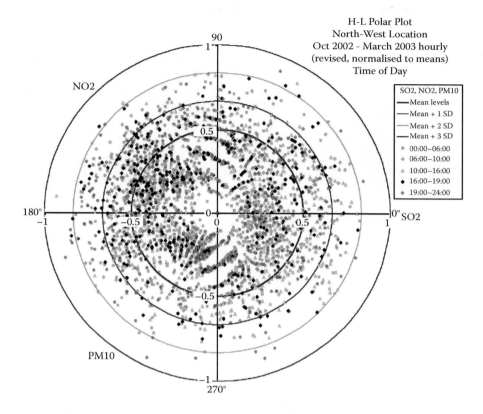

COLOR FIGURE 9.2.6
H-L polar diagram of SO$_2$, NO$_2$, PM10 airborne pollutants measured hourly over a six-month period, U.K. urban area. (H—pollutant; color of points—time of day; concentric circles (L)—mean level plus 1, 2, 3 standard deviations). (From Deakin, A. G., Rallis, I., Zhang, J., Spencer, J. W., and Jones, G. R. [2005]. Towards holistic chromatic intelligent monitoring of complex systems. *Proc. Complex Systems Monitoring Session, International Complexity, Science and Society Conference* (Liverpool). pp. 16–23. With permission.)

a)

b)

c)

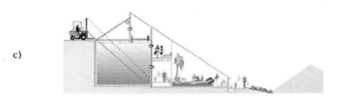

COLOR FIGURE 9.3.1
Experimental, industrial-scale anaerobic waste treatment unit. (a) View of the waste cells—greenhouse integrated unit, (b) vertical section of the unit showing the relative locations of the waste-processing cells and greenhouse, (c) filling a treatment cell with organic waste. (Courtesy AMEC. With permission.)

COLOR FIGURE 9.3.2
Chromatic monitoring instrumentation at the anaerobic treatment facility. (a) Pipe-work for recycling digestate, (b) CCTV remote monitor of leachate, (c) chromatic optical fiber pH and temperature sensors lance, (d) view of transparent pipe section and thermochromic elements. (Courtesy AMEC. With permission.)

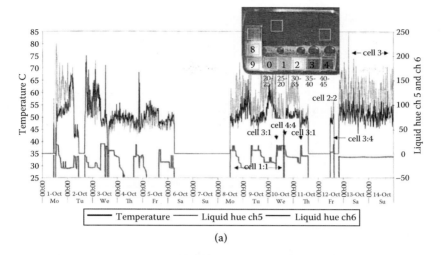

(a)

| 7:30 pm | 9:30 pm | 11:30 pm | 1:30 am | 3:30 pm |

(b)

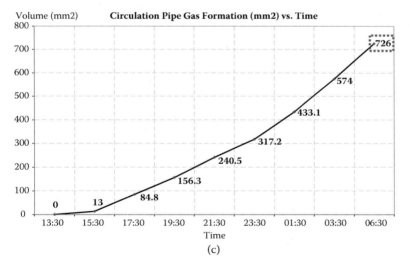

(c)

COLOR FIGURE 9.3.5

Time variation of conditions within the transparent length of the recirculating pipe. (a) Time variation of leachate temperature and chromaticity, (b) CCTV images of the pipe at various times showing biogas bubbles, (c) time variation of the volume of biogas bubbles determined using space chromatic processing of the CCTV images. ((b),(c) Rallis, I. (2004) Intelligent Chromatic Fiber Optic Sensors and Monitoring Systems for Enhancing Useful By-Products from Anaerobic Digestion. Ph.D. thesis, University of Liverpool; (a) Courtesy AMEC; Johnson, M. S. (2004) Research into the Anaerobic Biodigestion of Wastes, University of Liverpool Report.)

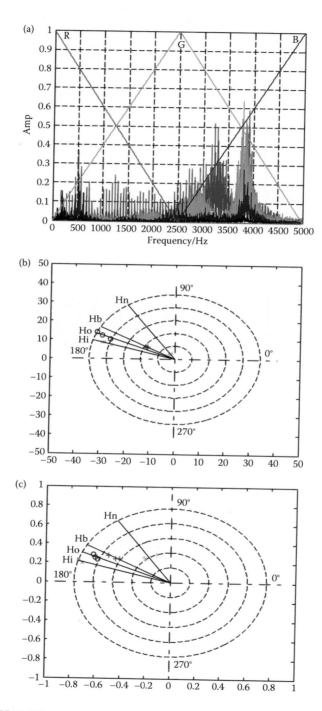

COLOR FIGURE 10.4.1
Frequency domain processing of acoustical signals from machine bearings. (a) Fourier-transformed signals showing R, G, B triangular filter superimposed. (b) H-L polar diagram, (c) H-S polar diagram (Faults: o—outer ring (Ho), x—inner ring (Hi), +—ball (Hb), *—normal (Hn)).

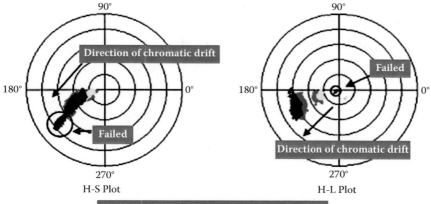

COLOR FIGURE 10.4.2
H-S, H-L polar diagrams for acoustic signals from a water pump at a combined heat and power plant recorded over a period of 5 months (color distinguishes calendar month). (From Cooke, R. (2000). An Intelligent Monitoring System, Ph.D. Thesis, University of Liverpool. With permission.)

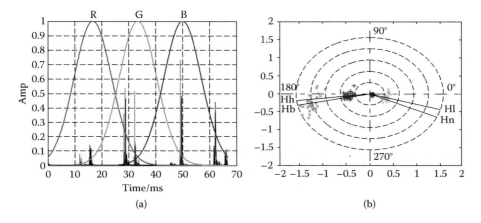

COLOR FIGURE 10.4.3
Fixed time domain processing of acoustical signals from a diesel engine. (a) One cycle of pulses with fixed time *R, G, B* Gaussian processors superimposed, (b) *H-L* polar diagram. (o—normal, x—blow by, +—high injection pressure, *—low injection pressure) (Engine data courtesy of Dr B. Liu, National University of Singapore. With permission.)

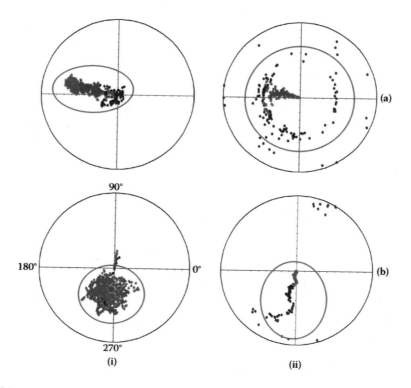

90°

180°   0°

270°

(i)   (ii)

(a)

(b)

COLOR FIGURE 10.4.8
Chromatic analysis results for the industrial rollbars. (a) A good-quality rollbar (b) a bad-quality rollbar. (i) *H-S*, (ii) *H-L*

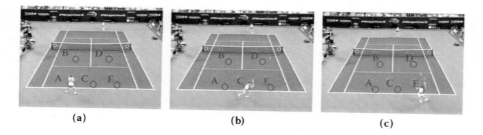

(a)   (b)   (c)

COLOR FIGURE 11.2.1.1
Chromatic position monitoring frames (a), (b), (c) show how a player changes the chromatic signature of monitoring points A, C, and E, respectively, when moving across the tennis court. (From Wong K., Xu S., and Jones G. R. (2005). Chromatic identification of complex movement patterns, *Proc. Int. Conf. on Complexity in Sci., Med. Sociol.*, Centre for Complexity Research, Liverpool. With Permission)

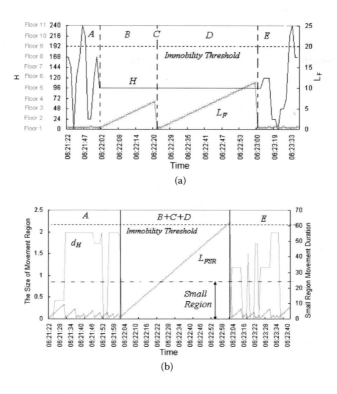

COLOR FIGURE 11.2.1.4
Time variation of parameters derived from chromatic processing. (a) Hs and L_F versus time—general movement, (b) movement size (d_H) and small-region-accommodated immobility (L_{FSR}) versus time. (From Xu S. and Jones G. R. (2006). Event and movement monitoring using chromatic methodologies, *IoP J. Meas. Sci. Tech.*, 17, 3204–3211. With Permission)

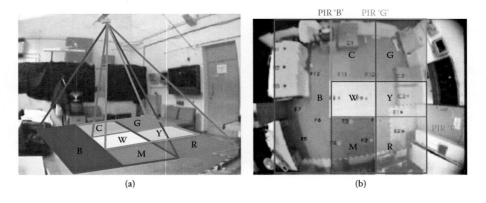

COLOR FIGURE 11.2.2.3
Deployment of PIR in trefoil space chromatic domain. (a) Isometric view. (b) Plan view.

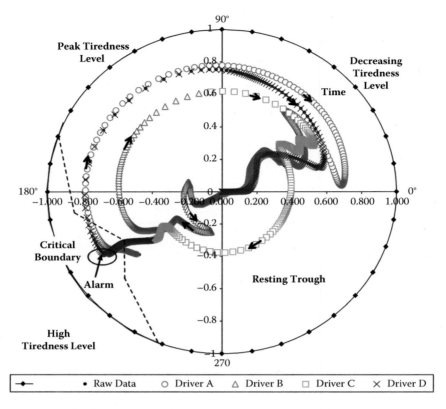

COLOR FIGURE 11.3.2.1
Chromatic polar diagram H-L for the overall output signal of Figure 11.3.1.2. (Different symbols corresponds to various drivers A–D, Figure 11.3.1.2) (From Koh A., Jones G. R., Spencer J. W., and Thomas I. (2007). Chromatic analysis of signals from a driver fatigue monitoring unit, *Meas. Sci. Technol.* 18, 1–8. With permission.)

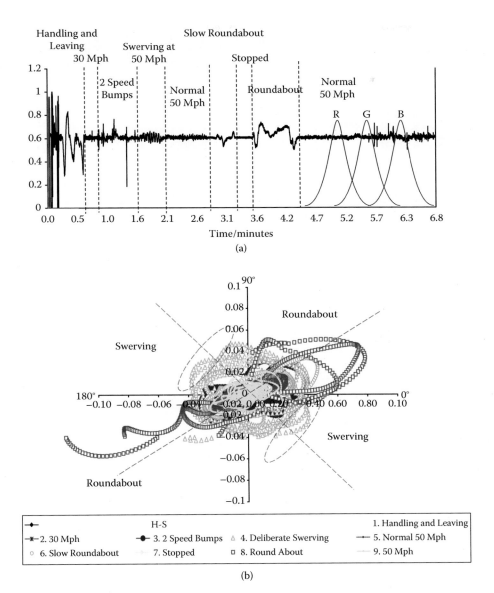

COLOR FIGURE 11.3.3.1
Chromatic processing of gyroscope signals. (a) Gyroscope output showing different signals associated with various driving features. (b) H-S polar diagram of the gyroscope output of Figure 11.3.3.1(a) (different symbols indicate different maneuvers). (From Koh A., Jones G. R., Spencer J. W., and Thomas I. (2007). Chromatic analysis of signals from a driver fatigue monitoring unit, *Meas. Sci. Technol.* 18, 1–8. With permission.)

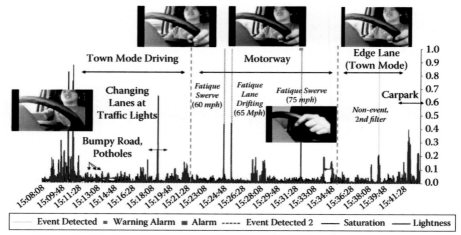

COLOR FIGURE 11.3.3.3
Real-time chromatic analysis of the gyroscope output.

(a)   (b)   (c)

FIGURE 2.2.1
(See color insert following page 18). Effect of changes in chromatic processors response profiles on image perception: (a) basic image; (b) G processor modified; (c) G processor modified further.

Quantification of the manner in which different profile forms of tristimulus processors (*R, G, B*) affect the *H, L, S* chromatic space is achieved by processing a series of monochromatic signals (*S* = 1; Appendix 1.I) sequentially as well subjecting the system to a fully saturated signal (*R = G = B*). These tests show that the monochromatic boundary in *H, L, S* chromatic space remains fixed as a circle independent of the response profiles of the chromatic processors. In addition, the achromatic (fully saturated) point (*S* = 0) remains at the center of this universal circle. However, the correspondence between the *H* and *S* scales to the scale of the monitored parameter is affected.

2.2.2   *H*: Parameter Value Characteristics for Different Processor Response Profiles

The effect of different processor response profiles on the monochromatic hue boundary scale may be illustrated by considering three cases (Dean, 2003):

(a) Three Gaussian processors overlapping at their half-height points (Figure 2.2.2a)

(b) Three Gaussian processors with only a small degree of overlap (Figure 2.2.2b)

(c) Truncated triangular filters overlapping at half height (Figure 2.2.2c)

The variation of *H* with frequency of a monochromatic signal for each of the three cases is shown in Figure 2.2.3a through 2.3c. Figure 2.2.3a corresponding to the half-height Gaussian processors (Figure 2.2.2a) shows:

- An *H* response is only obtained for the frequency range within which the processor responses overlap.

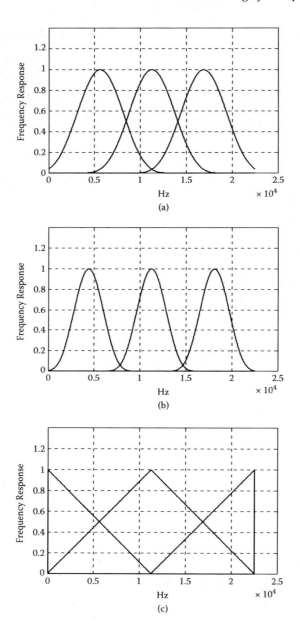

FIGURE 2.2.2
Various forms of chromatic processor profiles: (a) Gaussian processors, half-height overlaps; (b) Gaussian processors, marginal overlaps; (c) triangular processors, half-height overlaps. (From Koh, A., Dean, E.M., Zhang, J., Jones, G.R., and Spencer, J.W., *Proceedings of the Complex Systems Monitoring Session of the International Complexity, Science and Society Conference* [Liverpool], 2005. With permission.)

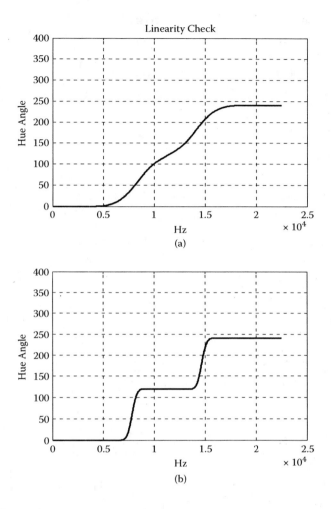

FIGURE 2.2.3
Hue:frequency relationships for monochromatic signals with the processor profiles of Figure 2.2.2: (a) Gaussian processors, half height overlaps; (b) Gaussian processors, marginal overlaps; (c) triangular processors, half height overlaps. (From Koh, A., Dean, E.M., Zhang, J., Jones, G.R., and Spencer, J.W., *Proceedings of the Complex Systems Monitoring Session of the International Complexity, Science and Society Conference* [Liverpool], 2005. With permission.)

- Within this range H varies monotonically between 0° and 270°.
- H variation is nonlinear with a reduced sensitivity mid range (R, B responses low) and maximum sensitivity at frequencies where the overlaps are substantial.

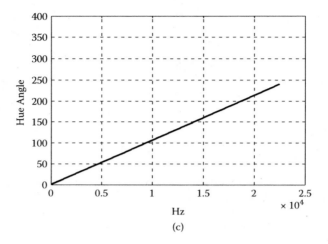

(c)

FIGURE 2.2.3
(Continued).

Figure 2.2.3b corresponding to the Gaussian filters with only a small degree of overlap (Figure 2.2.2b) shows:

- The H variation's nonlinearity is exaggerated.
- There is little sensitivity mid-range where there is no processor overlap but increased sensitivities in narrow regions on either side.

Figure 2.2.3c corresponding to the triangular filters with half-height overlaps (Figure 2.2.2c) shows that the H: frequency variation is linear throughout the range covered by the three processors.

The most linear characteristic occurred when the processors were separated by the width of the R and B processors.

The implications of these results are that the triangular filters provide a more uniform sensitivity throughout the frequency range but the Gaussian filters can provide regions of higher sensitivity, although they may not be so uniformly distributed. The triangular processors have a disadvantage in needing large kernels to accurately represent the discontinuities in the filter responses. However, they are ideal for processing discrete data sets where each data component is independent from each other (e.g., Section 2.5). For continuously distributed data the Gaussian processors (Figure 2.2.2a) are more practical.

2.2.3 Superposition of Monochromatic Signals

To gain an insight into the effect of processor characteristics upon S, three monochromatic signals of different frequencies and amplitudes may be superimposed and in turn with Gaussian and Triangular processors (Figure 2.2.4a,b) (Dean, 2003; Koh et al., 2005).

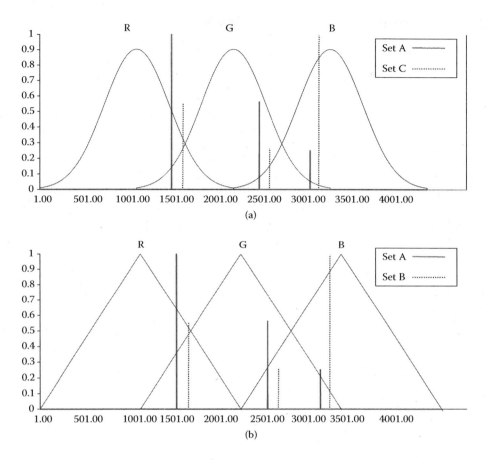

FIGURE 2.2.4
Superposition of three monochromatic signals: (a) Signals superimposed on Gaussian processors: (b) signals superimposed on triangular processors; (c) *H-S*, polar diagram showing signals A, B. (From Koh, A., Dean, E.M., Zhang, J., Jones, G.R., and Spencer, J.W., *Proceedings of the Complex Systems Monitoring Session of the International Complexity, Science and Society Conference* [Liverpool], 2005. With permission.)

Two cases of superimposed chromatic signals are considered corresponding to signals at each of three common frequencies but of different amplitudes (set *A*, *B* Figure 2.2.4a,b). The processed *H*, *S* values are shown on the *H-S* chromatic diagram of Figure 2.2.4c.

The difference between the *H* coordinate values for the Gaussian processed signals *A*, *B* ($240° - 64° = 176°$) is greater than for the triangular processed signals ($212° - 83° = 129°$). The difference in the *S* coordinates with the Gaussian processors is $0.46 - 0.39 = 0.07$ compared with $0.41 - 0.41 = 0$ with the triangular processors. For these particular processor profiles the Gaussian processors provide the better discrimination.

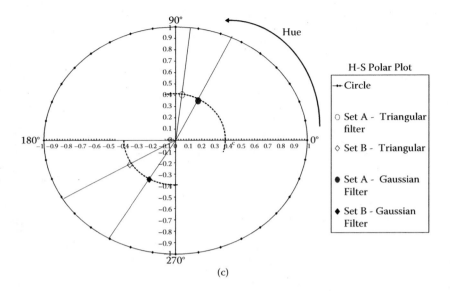

FIGURE 2.2.4
(Continued).

2.3   Multistimulus Chromatic Processing

2.3.1   Introduction

The chromatic methodology hitherto described has been based upon tristimulus processing. Stergioulas et al. (Stergioulas, 1997; Stergioulas et al., 2000) have shown that three chromatic processors are capable of discriminating approximately 95% of signals examined. For most practical monitoring purposes such a capability is acceptable but increasing the number of nonorthogonal processors can improve signal discrimination further (Stergioulas, 1997; Stergioulas et al., 2000). With six processors a signal discriminating capability of over 99% would seem possible. For practical purposes increasing the number of processors beyond six provides little gain and at the expense of added computational and practical sensing effort. Consequently, the use of six processors provides an upper limit to the number of processors needed for practical monitoring although three processors is optimum for most purposes.

There is a need to consider the implications of employing six rather than three nonorthogonal processors upon the chromatic transformation and the basic additional information produced.

2.3.2 Gaussian Families

The concept of a Gaussian function representing a family of signals processed by a given set of three nonorthogonal processors has been introduced in Chapter 1. This is based upon the assumption that the H, L, S values specific to a given signal can be regarded as representing the location of the Gaussian peak in parameter space, the signal strength (area under the Gaussian curve), and the width of the Gaussian function, respectively. Three such parameters are a necessary and sufficient condition to unambiguously define the specific Gaussian distribution.

Figure 2.3.1 shows an example to clarify the concept of a Gaussian family. Parts (i) and (ii) are two different signals but when processed with the same three nonorthogonal processors yield identical values of H, L, S. This indicates that although the signals are different in parameter space they may be regarded as belonging to the same Gaussian family in the chromatic space defined by the three specific processor characteristics with a mother Gaussian distribution shown as (iii) in Figure 2.3.1.

To place the discrimination capabilities in context, it needs to be recognized that human color vision, based upon a three-processor system, is capable of distinguishing in excess of 10^6 different colors. The monitoring capability of a tristimulus system may be regarded as similar and is consistent with the 95% signal discrimination quoted by Stergioulas et al. (2000).

An extension of chromatic processing for up to six nonorthogonal processors may be considered in terms of the mother Gaussian. For a truly Gaussian signal the use of a number of processors in excess of three would appear to be unnecessary as the three parameters of a tristimulus system is sufficient to define the Gaussian curve unambiguously. Thus, the use of up to six processors may be regarded as distinguishing between members of a single Gaussian curve family. For example, a skewed distribution might be distinguished from a Gaussian one using four processors; Kurtosis might be distinguished by five processors; six processors would be capable

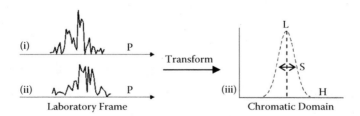

FIGURE 2.3.1
Equivalent Gaussian distributions. (From Jones G.R., Deakin, A.G., and Spencer, J.W., *Proceedings of the Complex Systems Monitoring Session of the International Complexity, Science and Society Conference* [Liverpool], 2005. With permission.)

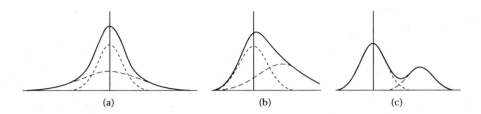

(a)   (b)   (c)

FIGURE 2.3.2
Signal distributions composed of two overlapping Gaussian distributions: (a) $L_G \neq L_X$, $S_G \neq S_X$, $H_G = H_X$ (Kurtosis); (b) $H_G \neq H_X$, S_G, $S_X \leq |H_G - H_X|$ (asymmetry); (c) $H_G \neq H_X$, S_G, $S_X >> |H_G - H_X|$ (resolved). (From Jones G.R., Deakin, A.G., and Spencer, J.W., *Proceedings of the Complex Systems Monitoring Session of the International Complexity, Science and Society Conference* [Liverpool], 2005. With permission.)

of distinguishing two superimposed Gaussian functions (Figure 2.3.2) (Rayleigh Criterion in spectroscopy).

2.3.3 Multistimulus Transformation

One method for transforming the outputs of more than three nonorthogonal processors into chromatic diagrams for convenient assimilation can be illustrated for the case of four processors. It is assumed:

- All processors are of the same form as each other, i.e., width, amplitude, profile.
- Three processors are identical to those used in the tristimulus system (i.e., *R*, *G*, *B*).
- The fourth processor overlaps another two processors symmetrically.

One form of processor deployment satisfying these conditions is shown in Figure 2.3.3a, *RGXB*.

The outputs of the four processors (*RGXB*) are separated into two sets *RGB* (equivalent to conventional tristimulus system) and *RXB* (equivalent to a skewed tristimulus system). Each of the two sets of tristimulus systems can then be transformed into its own *H*, *L*, *S* chromatic space (H_G L_G S_G and H_X L_X S_X). This may be regarded as providing a form of chromatic stereoscopic view of a signal being processed. That is, each tristimulus set assigns the signal to a particular Gaussian family within its chromatic space. Consequently, two different signals that belong to the same Gaussian family in one chromatic space may not belong to the same Gaussian family in the other chromatic space (e.g., Figure 2.3.3b). The two sets of *H*, *L*, *S* chromatic coordinates of the signal in each chromatic space may then be displayed together on single *H-S*, *H-L* polar diagrams (Figure 2.3.3c).

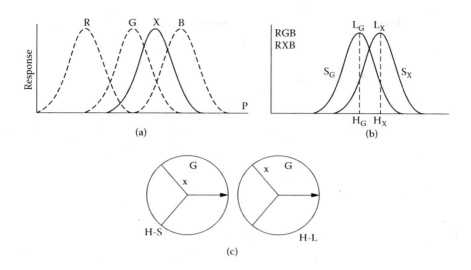

FIGURE 2.3.3
Four processor chromatic monitoring: (a) parameter space responsivities; (b) H_G L_G S_G and H_X L_X S_X Gaussian functions; (c) *H-S, H-L* diagrams showing the two equivalent Gaussian inputs X, G. (From Jones G.R., Deakin, A.G., and Spencer, J.W., *Proceedings of the Complex Systems Monitoring Session of the International Complexity, Science and Society Conference* [Liverpool], 2005. With permission.)

The process may be extrapolated to five or six processors, the addition of an extra processor leading to a further set of *H, L, S* coordinates (e.g., three with five processors).

2.3.4 Example of Deployment of Four-Processor System

An example of the deployment of a four-processor method is to discriminate between two signals that differ from each other but appear identical with tristimulus processing. For simplicity, the signals are each composed of the superposition of monochromatic signals. The first signal consists of the superposition of two monochromatic signals as shown in Figure 2.3.4a; the second consists of the superposition of three monochromatic signals as shown in Figure 2.3.4b. When addressed by three triangular processors *R, G, B*, the *H, L, S* coordinates of the two signals are identical as shown in the *H-S, H-L* polar diagram (Figure 2.3.4c). Consequently, the two signals belong to the same Gaussian family and are indistinguishable with the tristimulus processing.

When the two signals are addressed by the *R, X, B* tristimulus processors, their *H, L, S* coordinates differ as shown in the *H-S, H-L* polar diagram (Figure 2.3.4(d). The major difference occurs in the *S* coordinate value rather than *H* or *L*. Thus, in this particular case, it could be argued that it would be sufficient to use the fourth processor in conjunction with either *B* or *R* to simply calculate only the *BXR S* coordinate. The implication is that the fourth

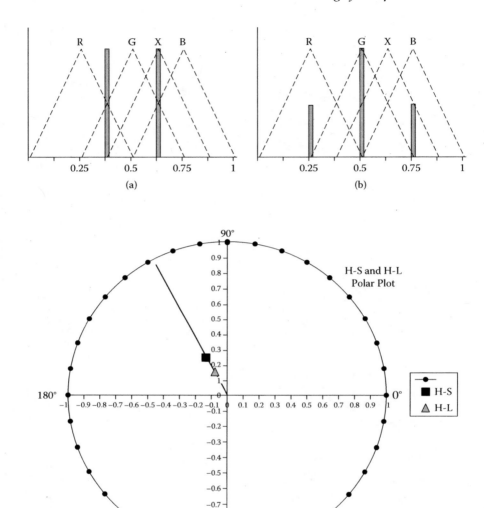

FIGURE 2.3.4

Four processor monitoring of superimposed monochromatic signals: (a) two monochromatic signals; (b) three monochromatic signals; (c) *H-S*, *H-L* diagram from *R*, *G*, *B* for signals (a) and (b) (*H-S*: ■ (a)=(b); *H-L*: ▲ (a)=(b)); (d) *H-S*, *H-L* diagram from *R*, *X*, *B* for signals (a) and (b) (*H-S*: ● (a), ◆ (b); *H-L*: ○ (a), ◊ (b)). (From Jones G.R., Deakin, A.G., and Spencer, J.W., *Proceedings of the Complex Systems Monitoring Session of the International Complexity, Science and Society Conference* [Liverpool], 2005. With permission.)

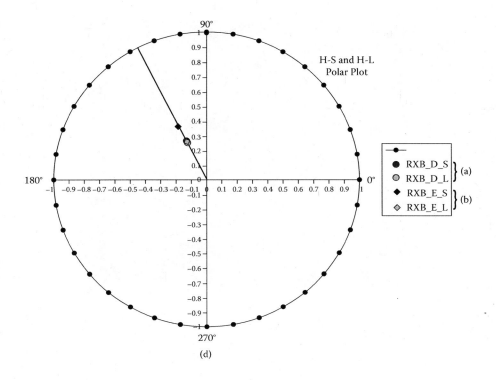

H-S and H-L
Polar Plot

RXB_D_S } (a)
RXB_D_L

RXB_E_S } (b)
RXB_E_L

(d)

FIGURE 2.3.4
(Continued).

processor differentiates further between various signal profiles produced by skewness or kurtosis.

2.4 Space-Domain Chromaticity

2.4.1 Introduction

The extension of the chromatic methodology to the spatial domain introduces additional considerations, given the 3-D nature of space. The space-based problems are those involving the locating of signal source and their distribution in two or three dimensions.

Simple examples of locating signal sources relate to identifying the location of personnel carrying such sources (tags), who operate in crowded environments (such as sporting events, Figure 2.4.1) or hazardous areas (e.g., high voltage substations). Sources without tags, such as warm objects or individuals may be located with pyroelectric infrared (PIR) detectors (Chapter 11, Section 11.2). The location and shape of a light emitting, filamentary, electric discharge may be monitored (Yokomizu et al., 1998; Chapter 4, Section 4.3).

FIGURE 2.4.1
Example of need for remote location determination in public areas (stewards at a crowded event). (With permission)

Locating entities emanating distinguishable signals is complicated by the nonuniform nature of the emanation in space, the nonlinear spatial distributions of detector responses, and the need for minimizing the number of sensing elements in the interests of economy, data acquisition, and efficiency.

Chromatic processing with a small number of remote, suitably deployed detectors having suitable spatial distributions of responsivity provides a convenient means for addressing such problems. Fundamental aspects of chromatic processing in 2-D space are considered in the following sections.

2.4.2   The Nature of Spatial Chromaticity

Space-based chromaticity refers to the deployment of chromatic monitoring for locating sources of signals in 2- or 3-D space. 2-D space deployment involves evaluating the two space coordinates—the radius (r) and azimuthal angle (θ). This may be achieved by deploying a cluster of three detectors/ processors in a circularly symmetric form within or around the space to be monitored. As such, it differs from the linear deployments of chromaticity described hitherto for wavelength, time, etc., applications (Chapter 1; Chapter 2, Section 2.2). In the linear case, it is only the central processor (G) (e.g., Figure 1.1b, Chapter 1; Figure 2.2.2, Chapter 2) that overlaps two other processors (R, B); in the circularly symmetric case, all the three processors' responsivities overlap. This in turn affects the chromatic H-L, H-S polar diagrams. There are two practically useful forms of circularly symmetric deployments: the trefoil and delta forms.

2.4.3   Circularly Symmetric Detectors

2.4.3.1   Trefoil Formation

A trefoil deployment of three detectors/processors (M_1, M_2, and M_3) is shown in Figure 2.4.2a whereby the detectors/processors are clustered together with their axes mutually inclined at 120°. Each detector/processor has an area of responsivity (a lobe) M_1, M_2, M_3 (Figure 2.4.2b) within which it responds to a signal to varying degrees according to the position of the signal source $V_{so}(r_v, \phi)$ relative to the detector as well as the source emission lobe.

Each processor lobe is normally defined in terms of the azimuthal angle (α, Figure 2.4.2b) with origin at the processor, and the extent of the lobe defines the area that can be monitored. The emission of the source $V_{so}(r_v, \phi)$ can be likewise defined.

In order to illustrate the adaptation of the chromatic approach to this situation, the responsivities of each processor (M_1, M_2, M_3) may be transformed from the coordinate system of each processor (α_1, α_2, α_3) to the coordinate system of the space being monitored i.e., r, θ (Figure 2.4.2a). Typical responsivities of such processors as a function of θ on a linear scale are as given in Figure 2.4.2c. This emphasizes the difference between the linear parameter (wavelength, etc.) cases and space monitoring. With the latter, the responsivities of each processor is nonorthogonal with respect to the other two processors rather than only the central processor (G) being so in the linear case.

The lobe of the source $V_{so}(r_v, \phi)$ can likewise be transformed into the r, θ coordinate system of the monitored space for each source location. A typical example of such a transformed signal is shown in Figure 2.4.2d for a source located at $\theta = 120°$. The manner in which the source distribution overlaps the response profile of each of the processors M_1, M_2, M_3 in this case is shown as the shaded areas.

The outputs (R, G, B) of detectors M_1, M_2, M_3 may then be processed chromatically using Equations 1.2 through 1.7 (Chapter 1) to yield values for the chromatic coordinates H, L, S. Some typical variations of H, L, S for various locations of a source are shown on the H-L, H-S polar diagrams (Figure 2.4.2e). The loci of a source moving through complete circuits (0–360°) around the detectors ($r = 0$) at two radii ($r_1 > r_2$) are shown. The forms of the loci reflect the variable sensitivities of the system with respect to H consistent with the H: parameter characteristic curve for the linear cases with Gaussian processor profiles presented in Figure 2.2.3a and 2.2.3b.

The relationships between the position coordinates r, θ and chromatic H, L, S are of the form:

$$r = f_1(H, S) \tag{2.4.1}$$

$$\theta = Hf(\phi, \alpha) \tag{2.4.2}$$

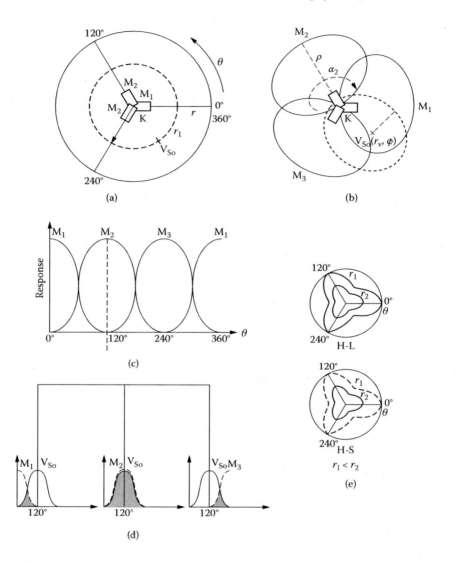

FIGURE 2.4.2
Remote chromatic monitoring of a signal source in 2-D space with a trefoil detection system:
(a) trefoil detection arrangement; (b) lobe diagram for the three detectors and signal source:
(c) representation of detector responses on a linear response: θ coordinate diagram; (d) signal
profile superimposed on the linear response: θ diagram; (e) H-L, H-S polar diagrams ($r_1 < r_2$).
(From Looe H.M., Lappas, C., Spencer, J.W., Jones, G.R., *Proceedings of the Complex Systems Monitoring Session of the International Complexity, Science and Society Conference* [Liverpool], 2005.
With permission.)

The precise forms of these relationships may be determined by theoretical calculation from the particular characteristics of the detectors and source or from *in situ* calibration tests.

2.4.3.2 Delta Formation

An alternative embodiment of circularly deployed processors is with the three processors M_1, M_2, M_3 in delta (Figure 2.4.3a) rather than in trefoil arrangement. In this case, the processor responses converge rather than diverge radially as in the trefoil case.

Figure 2.4.3a shows the deployment of the detectors in delta formation at the periphery of a monitored space along with their angular responses and

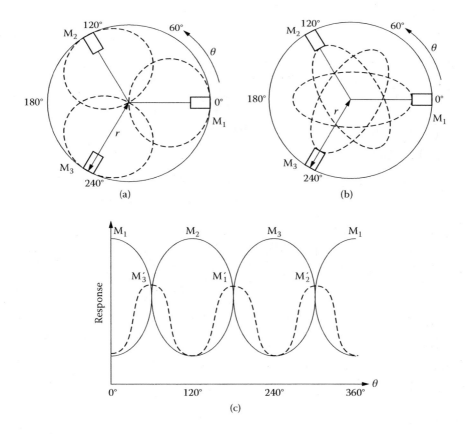

(a)

(b)

(c)

FIGURE 2.4.3
Remote monitoring with delta detection systems: (a) lobe diagram for detector range ~½ monitored range; (b) lobe diagram for detector range > ½ monitored range; (c) linear detector response: θ coordinate diagram for detectors with range >r. (From Looe H.M., Lappas, C., Spencer, J.W., Jones, G.R., *Proceedings of the Complex Systems Monitoring Session of the International Complexity, Science and Society Conference* [Liverpool], 2005. With permission.)

with a detection lobe of each processor approximately the radius of the circular environment.

The linearized responsivity-azimuthal angle (θ) diagram is of a similar form to the trefoil case (Figure 2.4.2c). This leads to H-S, H-L diagrams, which are also similar to those of the trefoil case (Figure 2.4.2e), except that the r_1, r_2 curves are reversed with the larger radius curve (r_2) being the outer curve. Values of r and θ are defined by Equations 2.4.1 and 2.4.2.

An example of a case whereby the detection lobe of each processor is greater than the radius of the monitored environment is shown in Figure 2.4.3b. The linearized responsivity-azimuthal angle (θ) diagram for this case (Figure 2.4.3c) exhibits additional satellite peaks produced by the extension of each detector lobe beyond the center of the monitored environment. For example, the main response of detector M_2 is at $\theta = 120°$, but there is a subsidiary response peak (M_2') at $\theta = 300°$. The resulting H-S, H-L characteristics become more complicated than for the trefoil case.

2.4.4   Conclusions

Chromatic methods may be employed for locating a source in 2-D space using weakly collimated sources and detectors.

Two basic processor arrangements are the trefoil and delta geometries.

The physical coordinates of a source (r, θ) in two dimensions are given by the chromatic parameters (H, S) and (H), respectively, Equations 2.4.1, 2.4.2. The third chromatic parameter, L, may be employed for ensuring correct system operation, e.g., absence of aberrations or noise.

The circular deployment of the processors with the trefoil or delta configurations leads to the responsivities of each processor overlapping with the responsivities of both the other processors, unlike the linear processor deployment when only one processor responsivity overlaps with two others. As a result Hue values in the range 270°–360° each represents a real location in physical space (unlike the linear parameter case, e.g., frequency, where it can only represent the superposition of two or more signals).

3-D space may be chromatically addressed by introducing additional processors (multistimulus processing) having 3-D response lobes to yield values for the three physical coordinates (r, θ, z).

2.5   Discrete Data Sets and Event Probability

2.5.1   Introduction

There are situations whereby the probability of an event occurring may depend, in a complex manner, upon a number of different parameters. The chromatic approach can be used for addressing the problem of indicating the probability of such an event in chromatic space rather than in terms of physical parameters.

(a)   (b)

FIGURE 2.5.1
(See color insert following page 18). Image of tree foliage: (a) spring/summer; (b) autumn. (Courtesy of Trevor Morris, Trevor Morris Photographics.)

An intuitive example of probability based upon chromatic indicators of a condition may be understood with the images shown in Figures 2.5.1a and 2.5.1b. These show images of the same tree at two different times of the year, one being predominantly green and the other orange. There is a high probability that the orange image (Figure 2.5.1b) indicates the time of the year to be autumn (although drought is not precluded), whereas the green image (Figure 2.5.1a) suggests springtime. There is a high probability that the change in color of the tree leaves indicates the tree to be deciduous.

The determination of the probability of an event using chromaticity often involves processing sets of discrete data rather than a continuous signal. The application of chromatic monitoring of such discrete data sets is first considered before describing the approach to determining event probability.

2.5.2 Discrete Data Processing

2.5.2.1 *Methodology*

Figure 2.5.2a shows how three nonorthogonal processors (R, G, B) may be applied to an array of discrete data ($D_1 - D_{10}$), for example, from the outputs of ten different sensors (Zhang, 2003; Zhang et al., 2005). In the case of discrete data processing, the profile of the chromatic processors can be advantageously triangular in form (Section 2.2). The output from each of the three processors is the sum of the products of the amplitude of each sensor output and the processor response to that sensor (Appendix 1.I).

$$R_0 = \sum_{n=1}^{10} [R(D) \cdot D_n] \qquad (2.5.1)$$

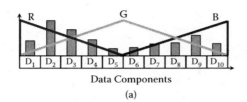

Data Components

(a)

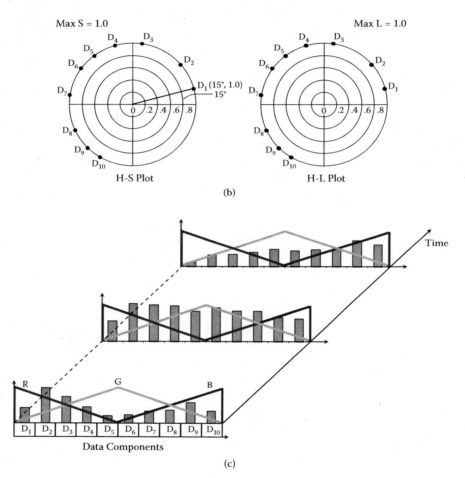

(c)

FIGURE 2.5.2
Chromatic processing of a discrete data set: (a) nonorthogonal processors applied to discrete data (D_1–D_{10}); (b) *H-S*, *H-L* polar diagrams for (D_1–D_{10}) data set; (c) nonorthogonal processors applied to time-varying data. (From Zhang J. and Jones, G.R., *Proceedings of the Complex Systems Monitoring Session of the International Complexity, Science and Society Conference* [Liverpool], 2005. With permission.)

D_n is the output of sensor n; $R(D)$ is the processor responsivity as a function of D.

The outputs of the three processors (R, G, B) are then mathematically transformed to yield the chromatic parameters H, L, S using Equations 1.2 through 1.7 in Chapter 1.

Figure 2.5.2b shows graphically, on H-S, H-L polar diagrams, the relationship between each sensor output and the hue angle H. This relationship is determined following the procedure described in Section 2.2 with respect to using monochromatic signals (Appendix 1.I) to calibrate H against a physical parameter (e.g., frequency). In the discrete data case, all sensor outputs apart from one are set to zero in turn, i.e.,

$$D_n = 1 \quad \text{otherwise} \quad D = 0 \tag{2.5.2}$$

Because in such a monochromatic case the saturation $S = 1$, this determines the location of the D_n sensor in terms of H on the circumference of the polar diagram. For example, for $D_n = D_1$, sensor D_1 output is located at $H = 15°$, $S = 1$, in Figure 2.5.2b. The location of the remaining sensor outputs, D_2–D_{10}, then follow as shown in Figure 2.5.2b.

The interpretation of H, L, S parameters when processing discrete data sets is as follows:

- H gives the dominant data component.
- L yields effective strength of the data set (i.e., the higher the integrated amplitudes of all signals, the larger is L).
- S is an indication of the spread of contributions between the members of the data set.

For example, a point with polar coordinates H, L, S having values (15°, 1, 1) indicates a single component ($S = 1$), which is D_1 ($H = 15°$) and high strength ($L = 1$). Coordinates (15°, 0.1, 0.1) indicate contributions from several datum points ($S = 0.1$) with low strength ($L = 0.1$) and with D_1 dominant.

2.5.2.2   Example of Implementation

An example of the manner in which the method may be implemented is for monitoring the relative extent to which various gases are produced by the degradation of insulating/cooling oil in extra high voltage power transformers (Zhang, 2003; Zhang and Jones, 2005). Each group of gases is indicative of a different degradation condition:

(a) CO, CO_2 > cellulose material degradation

(b) CH_4, C_2H_6, C_2H_4 > oil overheating

(c) C_2H_2, H_2 > electrical stressing

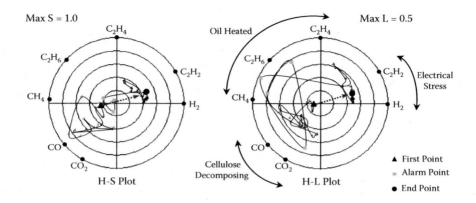

FIGURE 2.5.3

H-S, H-L polar diagrams for dissolved gas components for a high voltage transformer monitored over a 24-year period ($S = 0 \sim 1$, $L = 0 \sim 0.5$). (From Zhang J. and Jones, G.R., *Proceedings of the Complex Systems Monitoring Session of the International Complexity, Science and Society Conference* [Liverpool], 2005. With permission.)

The gases may be ordered according to these three groups, overlaid by three chromatic processors (R, G, B) and corresponding H, L, S values calculated and displayed on H-S, H-L polar diagrams. Various values of H then correspond to different groups. For example, the material decomposition group (CO_2, CO) correspond to 270–180° (D_7–D_6), the overheating group to 180°–90° (D_5–D_3) and the electrical stressing group to 90°–0° (D_2–D_1) (Figure 2.5.3).

By evaluating the H, L, S coordinates for the gaseous composition at different times (Figure 2.5.2c) time trajectories of the evolution of various potential fault conditions can be displayed on H-S, H-L diagrams (Figure 2.5.3). Alarms can be raised when certain levels are reached within the various H. Figure 2.5.3 shows how the state of a transformer progressed over two decades terminating in an indication of electrical discharging following periods of decomposition and overheating.

The extent of data compression is substantial without loss of significant information.

2.5.3 Probability of an Event

The likelihood of a system condition evolving may be indicated quantitatively in terms of the chromatic domain parameters H, L, S. The method may be illustrated using data from Figure 2.5.3 from the power transformer tests already presented in Section 2.5.2 (Zhang, 2003; Zhang et al., 2005).

The results of these tests indicate the following:

- Copious gas production (high values of L) suggest an increased probability $P(L)$ of fault development.

- A predominance of high-risk components (H_2, C_2H_2) signifies an increased risk $P(H)$ of transformer failure.
- An enhancement of one group of gases compared with others (high S) signifies an increased probability $P(S)$ of an undesirable condition evolving.

The hierarchical order of the H, L, S parameters for indicating fault probability as determined from 50 transformer tests was $L>H>S$. The overall probability of a fault may then be expressed as the sum of the L, H, S probabilities, each of which may be assumed to be a Gaussian function (Papoulis and Pillai, 2002).

$$P(L,H,S) = l \cdot P_L + h \cdot P_H + s \cdot P_S = P_0 \left\{ l \cdot \exp\left[-\frac{1}{2}\left(\frac{L-L_m}{\sigma_L} \right)^2 \right] \right.$$

$$\left. + h \cdot \exp\left[-\frac{1}{2}\left(\frac{H-H_m}{\sigma_H} \right)^2 \right] + s \cdot \exp\left[-\frac{1}{2}\left(\frac{S-S_m}{\sigma_S} \right)^2 \right] \right\}$$

(2.5.3)

where l, h, s are scaling factors with

$$l + h + s = 1 \qquad (2.5.4)$$

and

$$L_m = 1, \quad S_m = 1, \quad H_m = 240°.$$

For the transformer case estimated values of l, h, s were $l = 0.5$, $h = 0.3$, $s = 0.2$.

The probability of a fault can be updated each time a gas analysis is made and chromatically processed (Figure 2.5.2c) so that a time-based trend in fault probability can be produced to check whether it is increasing and if so at what rate.

2.6 Reorganization of Complex Data Sets

2.6.1 Introduction

Sets of discrete data obtained from tests affected by many operating conditions that are not easily controlled can contain substantial amounts, levels, and types of information. However, it is often difficult to identify, extract, and quantify significant parts of the information when there is no prior knowledge.

Such data may be addressed by extending the chromatic approach to discrete data processing (Section 2.5) by rearranging the order of the data elements (D_1–D_{10}, Figure 2.5.2a) in the set along the parameter axis. As a result, different types of information emerge, which can be displayed and quantified on chromatic polar diagrams.

To illustrate such possibilities, data that have been gathered about airborne microparticulates over several weeks at many sites on different days, various time of day, and under many weather conditions can be used (Reichelt et al., 2006; Kolupula et al., 2005; Chapter 9). Being real sites, the operating conditions (e.g., vehicle traffic density) could not be controlled or could the complex manner in which they might interact.

2.6.2   Data Complexity and Processing

Figure 2.6.1 shows airborne particle concentrations determined from a series of tests (Reichelt et al., 2006) under the following conditions:

(a) At six different locations (A–F)

(b) On ten different occasions

(c) Two different days of the week (Tu, Th)

(d) Two periods during each day (early, late p.m.)

(e) During two different months (N, D)

(f) Various humidity levels (high, H; medium, M; low, L)

Histograms of particle concentrations at each location over the duration of the tests are shown for the late and early afternoon periods in Figure 2.6.1a and 2.6.1b, respectively. An expanded view of the ten-day results at one particular location is shown in Figure 2.6.1c whereas the day, humidity level, and month are identified in Figure 2.6.1d.

Thus, the raw data incorporates the possible influence of a number of complex and interactive effects about which information might be sought.

The discrete nature of the data allows the chromatic processing described in Section 2.5 to be applied. For example, three triangular chromatic processors (R, G, B) may be overlaid on to the particular concentration histogram of the type shown in Figure 2.6.1c. The R, G, B outputs are converted into H, L, S chromatic parameters, which can be displayed on chromatic polar diagrams.

There are several ways in which the discrete data given in Figure 2.6.1a and 2.6.1b may be grouped to form sets of the types appearing in Figure 2.6.1c ready for chromatic processing. The grouping needs to emphasize the type of information being sought. For example, if information about the distribution of particulate levels at a location throughout the entire monitoring period during the early afternoon is sought, the discrete data needs to be ordered sequentially according to calendar dates, as shown in Figure 2.6.1b. The procedure can be applied to each location in turn and also to the late afternoon period.

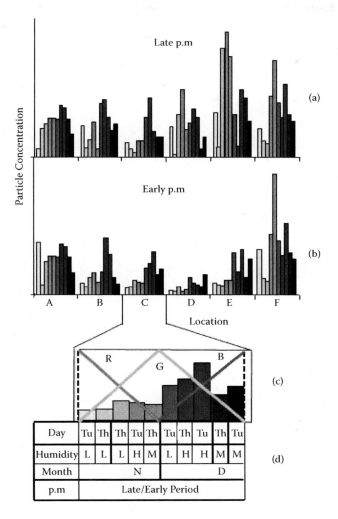

FIGURE 2.6.1

Airborne particles discrete data: (a) atmospheric particulates at six locations on ten different days—late p.m.; (b) atmospheric particulates at six locations on ten different days—early p.m.; (c) expanded view of ten-day results at one location; (d) additional embedded information (day of week, humidity, month, part of day). (From Kolupula, Y.R., Reichelt, T.E., Aceves-Fernandez, M.A., Deakin, A.G., Spencer, J.W., and Jones, G.R., *Proceedings of the Complex Systems Monitoring Session of the International Complexity, Science and Society Conference* [Liverpool], 2005. With permission.)

If alternate information is sought about the distribution of particulate levels throughout the monitoring period on each type of day (e.g., Tuesday [Tu]) during the early afternoon and with less emphasis on location, the discrete data from each location may be divided into groups corresponding to each day type (Figure 2.6.2).

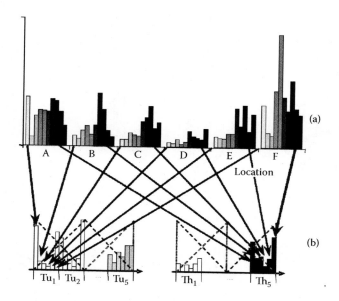

FIGURE 2.6.2
Data reorganization for "Day of the Week Information": (a) raw location-based data; (b) day-based reorganized data. (From Kolupula, Y.R., Reichelt, T.E., Aceves-Fernandez, M.A., Deakin, A.G., Spencer, J.W., and Jones, G.R., *Proceedings of the Complex Systems Monitoring Session of the International Complexity, Science and Society Conference* [Liverpool], 2005. With permission.)

Other groupings can be organized to yield further information (e.g., the variation of particulate levels with humidity).

Since chromatically processing any such set compresses the data into three coordinate values (H, L, S), a range of situations may be conveniently compared on a single H-S polar diagram. For example, results for the particulate distribution throughout the monitored period for each of the six locations (A–F) may be compared on a single H-S polar diagram. Similarly, the particulate distribution on each day type (Tu, Th) and time of the day (early, late afternoon) may be compared on another H-S polar diagram.

2.6.3 Examples of Information Emphasis

The manner in which the chromatic processing of the variously grouped discrete data sets emphasizes different information on an H-S polar diagram is illustrated with the two examples given in Section 2.6.1:

(a) Sequential ordering according to calendar data of individual locations (Figure 2.6.1c)

(b) Sequential ordering according to calendar date of a single type of day (Tu or Th) but at all locations collectively (Figure 2.6.2b)

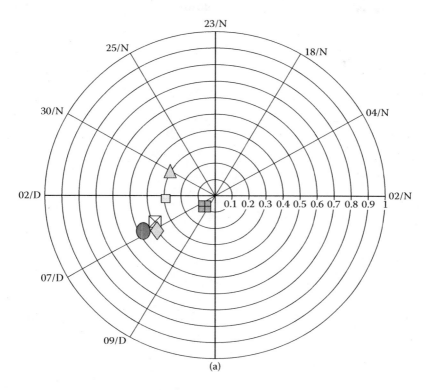

(a)

FIGURE 2.6.3
H-S(L) Polar diagrams for accumulated particle levels: (a) distribution during monitoring period clearly p.m.; (Location A—⊞, B—⊠, C—○, D—□, E—◇, F—△); (b) distributions on two different days of the week (early and late p.m.) (Tu early—⊞, late—□, Th early—◇, late—△). (From Kolupula, Y.R., Reichelt, T.E., Aceves-Fernandez, M.A., Deakin, A.G., Spencer, J.W., and Jones, G.R., *Proceedings of the Complex Systems Monitoring Session of the International Complexity, Science and Society Conference* [Liverpool], 2005. With permission.)

H-S polar diagrams based upon the particulate levels data of Figure 2.6.1a for these two cases are given in Figure 2.6.3a and 2.6.3b.

On these diagrams the azimuth angle represents *H* and the radius *S*. *L* is represented in five levels (0–0.2 to 0.8–1.0) by gradations of a grey code (0–0.2 white to 0.8–1.0 black) of the point representing a condition on the *H-S* diagram. A further piece of information (e.g., location, day type) is represented by the symbol representing the data point.

Thus, a data point carries four sets of quantized information *H*, *L*, *S* and symbol type so that several points representing different situations can be unambiguously shown for comparison on a single *H-S(L)* polar diagram.

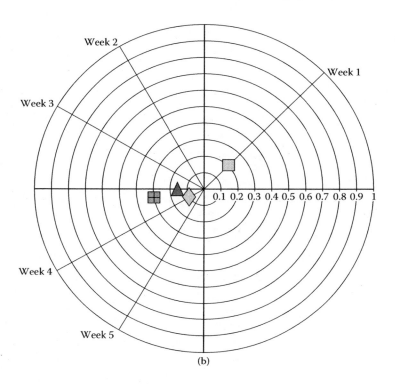

(b)

FIGURE 2.6.3
(Continued).

2.7 Summary

The effect of changing the chromatic processor characteristics and the add-ressing of discrete data sets have been described.

The *H, L, S* monochromatic boundary and achromatic point remain fixed but the scaling between physical parameter and chromatic domains is affected by changing the chromatic processor characteristics.

Triangular processor profiles provide linear *H*-physical parameter rela-tionships with monochromatic signals but can lead to truncation difficulties at discontinuities in the profiles.

Gaussian processor profiles provide sensitivities that vary throughout the physical parameter range so providing a capability to tune chromatic transformations.

The use of three chromatic processors (tristimulus) is optimum for most practical monitoring applications.

Increasing the number of chromatic processors beyond six does not pro-vide proportionate, additional information and requires additional hard-ware and software.

Application of chromatic monitoring in the space domain with circularly deployed processors whose angular responses are variable, produces an equal degree of nonorthogonality between all chromatic processors.

The introduction of a fourth processor and 3-D angular responses for each processor allows the location of a source to be determined in 3-D space.

Discrete data sets can be advantageously addressed by triangular chromatic processors as truncation difficulties are more easily accommodated.

Discrete chromatic processing can be utilized for extracting and quantifying information from the outputs of a series of different sensors addressing various parameters.

The probability of events occurring or conditions evolving can be quantified in terms of chromatic rather than physical parameters, which may be regarded as representing an overall system behavior.

Chromatic processing of discrete data sets obtained from complex conditions can be deployed for extracting different forms of information by reorganizing the ordering of members of the data sets with respect to the chromatic processor profiles.

Different information may be emphasized and displayed on a single *H-S(L)* diagram.

References

Dean, E.M. (2003). Nonintrusive passive acoustic monitoring of liquid flow systems. Ph.D. thesis, University of Liverpool.

Jones, G.R., Deakin, A.G., and Spencer, J.W. (2005). Multistimulus chromatic processing of complex systems, *Proceedings of the Complex Systems Monitoring Session of the International Complexity, Science and Society Conference* (Liverpool), pp. 5–15.

Koh, A., Dean, E.M., Zhang, J., Jones, G.R., and Spencer, J.W. (2005). Effect of chromatic filter characteristics in quantifying complex data, *Proceedings of the Complex Systems Monitoring Session of the International Complexity, Science and Society Conference* (Liverpool), pp. 46–51.

Kolupula, Y.R., Reichelt, T.E., Aceves-Fernandez, M.A., Deakin, A.G., Spencer, J.W., and Jones, G.R. (2005). Chromatic methodologies for information extraction from complex data sets, *Proceedings of the Complex Systems Monitoring Session of the International Complexity, Science and Society Conference* (Liverpool), pp. 68–76.

Looe, H.M., Lappas, C., Spencer, J.W., Jones, G.R. (2005). Location determination of an entity by remote chromatic processing of its emanations, *Proceedings of the Complex Systems Monitoring Session of the International Complexity, Science and Society Conference*, pp. 29–34.

Papoulis, A. and Pillai, S.U. (2002). *Probability, Random Variables, and Stochastic Processes*, 4th ed., McGraw-Hill, New York.

Reichelt, T.E., Aceves-Fernandez, M.A., Kolupula, Y.R., Pate, A., Spencer, J.W., and Jones, G.R. (2006). Chromatic modulation monitoring of airborne particulates, *Meas. Sci. Technol.* 17, 675–683.

Stergioulas, L.K. (1997). Time-frequency methods in optical signal processing, Ph.D. thesis, University of Liverpool.

Stergioulas, L.K., Vourdas, A., and Jones, G.R. (2000). Gabor representation of a signal using a truncated von Neumann lattice and its practical implementation, *Opt. Eng.*, 39(7), 1965–1971.

Yokomizu, Y., Spencer, J.W., and Jones, G.R. (1998). Position location of a filamentary arc using a tristimulus chromatic technique, *J. Phy. D: App. Phy.*, 31(23), pp. 3373–3382.

Zhang, J. (2003). Chromatic identification of incipient transformer failures due to partial discharges, Ph.D. thesis, University of Liverpool, U.K.

Zhang, J. and Jones, G.R. (2005). Chromatic processing for event probability description, *Proceedings of the Complex Systems Monitoring Session of the International Complexity, Science and Society Conference* (Liverpool), pp. 61–67.

Zhang, J., Jones, G.R., Deakin, A.G., and Spencer, J.W. (2005). Chromatic processing of DGA data produced by partial discharges for the prognosis of HV transformer behaviour, *Meas. Sci. Technol.*, 16(2), pp. 556–561.

3

Other Chromatic Processing Algorithms

G.R. Jones, P. Pavlova, and J.W. Spencer

CONTENTS

3.1 Introduction

Although the basis of chromatic methods can be related to the photic field concepts of Moon and Spencer (1981) and Gabor transforms (Jones et al., 2000), the adaptation of the *H, L, S* algorithms of color science provide a convenient basis for quantifying monitored trends in terms of signal features such as dominant frequency, signal strength, and bandwidth. However, as the perception of color by the human vision system is one particular form of general chromaticity, various algorithms employed in color science, other than the *H, L, S* formulations, can in principle be broadened and adapted for chromatic monitoring and processing.

Such approaches can offer practical advantages in terms of the type and number of chromatic processors deployed as well as computational advantages in terms of processing times and software simplicity. Several such algorithms other than *H, L, S* exist. Three examples used for chromatic monitoring are described: the *H, V, S* (e.g., Rogers, 1985), the *x, y, z* (e.g., Jones and Russell, 1993), and the *Lab* (Schwarz et al, 1987) schemes, in relation to practical chromatic monitoring.

3.2   The *H, S, V* Transformation

The *H, S, V* transformation algorithms are similar to the *H, L, S* algorithms of Chapter 1 but provide a different emphasis on certain signal features. In this case, the outputs from the *R, G, B* processors (Appendix 1.I) are transformed into *H, S, V* parameters with the following algorithms (Rogers, 1985):

$$V = \max (R, G, B) \tag{3.1}$$

$$S = \frac{V - \min}{V} \tag{3.2}$$

where min = min (*R, G, B*),
   H is evaluated from the following relationships between *R, G, B*, and *V*

$$c_r = \frac{V - R}{V - \min}; \quad c_g = \frac{V - G}{V - \min}; \quad c_b = \frac{V - B}{V - \min}$$

if $V = R$ $H = c_b - c_g$

if $V = G$ $H = 2 + c_r - c_b$

if $V = B$ $H = 4 + c_g - c_r$

$$H = H * 60 \quad \text{and} \quad \text{if} \quad H < 0 \quad H = H + 360 \tag{3.3}$$

Thus, *V* is the value of the highest output from the three processors, *S* is the saturation relative the highest processor output, *H* represents the dominant wavelength. The *H, S, V* parameters may be represented graphically in a similar manner to *H, S, L* (Figure 1.4a) on a *θ, r, z* polar diagram. In the *H, S, V* case the polar diagram is in the form of an inverted cone (Figure 3.1) whose apex corresponds to low signal levels ("black") and whose base represents maximum signal levels (Schwarz et al., 1987; Aïnouz et al., 2006).

   From the aspect of color science, the *H, S, V* transformation provides a model which is more akin to human color vision than *H, S, L*. From the aspect of signal discrimination, it provides a better *S* resolution than the *H, L, S* model, which can be important when discriminating between signals which predominantly differ by their saturation levels.

   The *H, S, V* algorithms have been applied to differential blood counting (Pavlova et al., 1996; Chapter 8, Section 8.5).

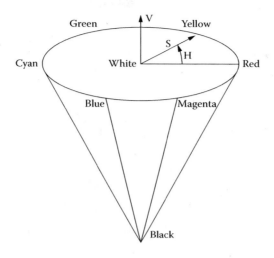

FIGURE 3.1
Graphical representation of the *HSV* model (Wikipedia).

3.3 The *x, y, z* Transformation

3.3.1 Basic *x, y, z* Algorithms

Another transformation used in color science, which is adaptable for chromatic monitoring and processing, is the *x, y, z* system, on which the CIE diagram of color science is based (e.g., Billmeyer and Saltzman, 1981). The *x, y, z* parameters are defined in terms of the outputs of the three chromatic processors *R, G, B* (Appendix 1.I) by Equations 3.4 through 3.6 (e.g., Jones and Russell, 1993).

$$x = R/(R + G + B) \tag{3.4}$$

$$y = G/(R + G + B) \tag{3.5}$$

$$z = B/(R + G + B) \tag{3.6}$$

so that

$$x + y + z = 1 \tag{3.7}$$

The CIE diagram is a Cartesian representation in terms of *x:y* (Figure 3.2a) with *z* being degenerate because of the relationship (3.7).

A further parameter *L* as defined by Equation 1.3 (Chapter 1) is also relevant to indicate the lightness (i.e., the signal strength). From a monitoring perspective, *x* and *y* represent the relative contributions of two parts of a spectrum.

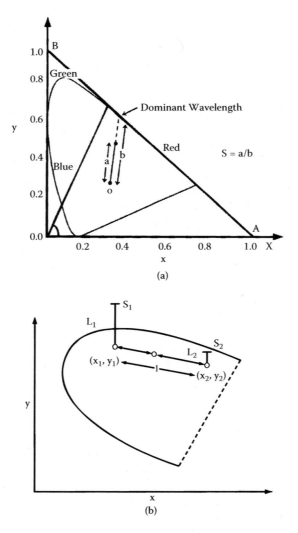

FIGURE 3.2

Chromatic diagrams *x:y*: (a) manifestation of dominant wavelength and saturation (S); (b) combination oft two signals (L_1, x_1, y_1), (L_2, x_2, y_2). (From Jones, G.R. and Russell, P.C., *Pure Appl. Opt.*, 2, 87–110, 1993. With permission.)

As with the *H, L, S* system, monochromatic signals lie on the periphery of the color space defining curve. The point corresponding to $R = G = B$ (i.e., $S = 0$ in *H, L, S* space) has coordinates (0.33, 0.33). The dominant wavelength of a color is determined by the intersection of the line from (0.33, 0.33) through the point representing the signal, extrapolated to intersect the color space boundary (*ob*, Figure 3.2a). Its saturation is given by *a/b* (Figure 3.2a).

From a monitoring perspective, an advantage of the x, y, L system is that it enables the result of superimposing two signals upon each other to be conveniently quantified and displayed. For example, the resultant chromaticity of two superimposed signals, (x_1, y_1, L_1) and (x_2, y_2, L_2) (Figure 3.2b) can be conveniently calculated with the algorithms of Equations 3.8 through 3.10 (e.g., Jones and Russell, 1993).

$$x_3 = x_1 \left(L_1 \Big/ (L_1 + L_2) \right) + x_2 \left(L_2 \Big/ (L_1 + L_2) \right) \tag{3.8}$$

$$y_3 = y_1 \left(L_1 \Big/ (L_1 + L_2) \right) + y_2 \left(L_2 \Big/ (L_1 + L_2) \right) \tag{3.9}$$

$$L = R + G + B \tag{3.10}$$

In relation to chromatic monitoring, the x, y, z algorithms (3.4), (3.5), (3.6) of color science may be regarded as a particular example of a more general transformation when the constraint of absolute color definition for human vision is relaxed.

3.3.2 Generalized Chromatic Algorithms Based Upon *x, y, z*

More generalized forms (x_0, y_0) of the x, y algorithms (3.4), (3.5) may be defined as follows:

$$x_G = \frac{A[a\,R_0 - (m_1\,R_0 + m_2\,G_0 + m_3\,B_0)]}{[n_1\,R_0 + n_2\,G_0 + n_3\,B_0]} \tag{3.11}$$

$$y_G = \frac{B[b\,G_0 - (m_1\,R_0 + m_2\,G_0\,m_3\,B_0)]}{[n_1\,R_0 + n_2\,G_0 + n_3\,B_0]} \tag{3.12}$$

where A, B, a, b, m_1, m_2, m_3, n_1, n_2, n_3 are weighting factors that can be controlled either in software or hardware (e.g., amplifier gains). In this respect the x, y, z algorithms can be more convenient for software or hardware implementation than the H parameter of the H, L, S transform.

There are a number of particular cases of Equations 3.11 and 3.12, which have been deployed with chromatic monitoring for various applications.

(a) *Distimulus Monitoring*

Distimulus monitoring involves only two nonorthogonal processors. This corresponds to $A = B = a = b = n_1 = n_2 = 1$, $m_1 = m_2 = m_3 = n_3 = 0$. in Equations 3.11 and 3.12 so that

$$x = R_0/(R_0 + G_0) \tag{3.13}$$

$$y = G_0/(R_0 + G_0) \tag{3.14}$$

Equations 3.13 and 3.14 enable the instrumentation to be simplified and have been used in cases wherein optical spectral changes are progressive (e.g., Murphy and Jones, 1993). In this case the x:y variation is monotonic (line AB Figure 3.2a). In this case the dominant wavelength is given quite simply by (Jones and Russell, 1993):

$$\lambda_d = \tan \theta = x/y = R_0/G_0 \tag{3.15}$$

(b)   Tuned Distimulus Monitoring

Although only two processors are employed in tuned distimulus monitoring, there is flexibility in dealing with their outputs in that A, B, a, b, m_1, m_2, n_1, n_2 are not restricted to unity but $n_3 = m_3 = 0$. Equations 3.11 and 3.12 are reduced to

$$x = \frac{A[(a - m_1)R_0 - m_2 G_0]}{[n_1 R_0 + n_2 G_0]} \tag{3.16}$$

$$y = \frac{B[(b - m_2)G_0 - m_1 R_0]}{[n_1 R_0 + n_2 G_0]} \tag{3.17}$$

These two equations lead to a relationship for the dominant wavelength of the form (e.g., Li et al., 1999; Jones et al., 1998).

$$\lambda_d' = x/y = k_0[R_0 - k_1 G_0]/[R_0 + k_2 G_0] \tag{3.18}$$

where k_0, k_1, k_2^* are weighting factors related to A, B, a, b, m_1, m_2. The graphical form of the relationship (3.18) shown in Figure 3.3 is a sigmoid curve, whose location on the dominant wavelength axis (λ_d') and its inclination can be adjusted via the scaling factors k_1, k_2. This enables the characteristic to be tuned to the appropriate dominant wavelength range and sensitivity.

Equation 3.18 has an additional significance in that for $k_0 = k_1 = k_2 = 1$ it reduces to

$$\lambda_d' = (R_0 - G_0)/(R_0 + G_0) \tag{3.19}$$

which is of the same mathematical form as the equation defining the saturation, S, in the H, L, S chromatic transform (Equation 1.4, Chapter 1).

Equation 3.18 is the form used by Jones et al. (1998) for monitoring electric current via the magneto-optic effect (Chapter 6, Section 6.3.5).

(c)   Basic Tristimulus Processing

Basic tristimulus processing involves three processes with $A = B = n_1 = n_2 = n_3 = a = b = 1$ and $m_1 = m_2 = m_3 = 0$ reducing Equations 3.11 and 3.12 to Equations 3.4 and 3.5 of the CIE x:y diagrams.

---

* $k_0 = (A/B)(1 - a/m_1)$; $k_1 = m_2/(a - m_1)$; $k_2 = (m_2 - b)/m_1$ from Equations 3.11, 3.12, and 3.18.

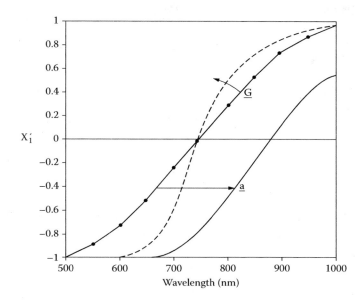

FIGURE 3.3
Distimulus processing characteristics. $X_i' = \lambda_d' \frac{K_0 (R_0 - K_1 G_0)}{(R_0 + K_2 G_0)}$, $(-1 < X_i' < 1)$; $a = K_1$, $G = K_0$, are chosen to zoom on a particular region.

However, for chromatic monitoring purposes the responsivities of the R, G, B processors are not restricted to those of the receptors of the human eye but may be varied to suit a particular application or available hardware detectors and sources.

The effect of varying such responses is to modify the shape of the chromatic space (Jones and Russell, 1993). This may be illustrated by considering five different processor responses ($R_1 - R_5$) such as those shown in Figure 3.4.

The monochromatic boundaries of $x{:}y$ chromatic diagrams based on three permutations of three of these profiles (e.g., [R_1, R_2, R_3], [R_1, R_3, R_4], [R_3, R_4, R_5]) can be determined by examining the $x{:}y$ coordinate values derived for a series of monochromatic signals at different wavelengths/frequencies. The achromatic point can be determined by making the outputs from the three processors equal.

The resulting $x{:}y$ chromatic space for these three combinations of processor profiles are shown in Figures 3.5a through 3.5c. The achromatic point remains fixed with coordinates (0.33, 0.33), but substantial differences are produced in the shape of the monochromatic boundary by such changes in the profiles of the processors. Consequently, significant changes in chromatic modulation depth (up to 4.5x (Jones and Russell, 1993)) can be produced by the choice of appropriate processor combinations. Such considerations need to be taken into account when choosing appropriate processors for various types of optical fiber sensors, for example, ones based upon thin films or optically active materials (Chapter 6).

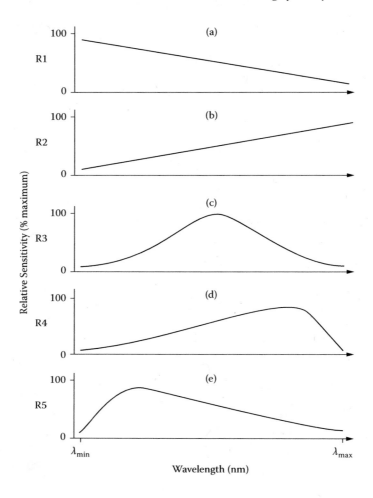

FIGURE 3.4
Responsivities of various types of chromatic processors: (a) R_1; (b) R_2; (c) R_3; (d) R_4; (e) R_5. (From Jones, G.R. and Russell, P.C., *Pure Appl. Opt.*, 2, 87–110, 1993. With permission.)

An advantage of the tristimulus monitoring is that it enables trend patterns involving two variables to be obtained for further analysis.

(d) Tristimulus Diagrams with Modified Coordinates

There are cases whereby the tristimulus processor outputs may be advantageously displayed upon chromatic maps that have a different coordinate system from the *x:y* CIE type diagram. An example of this is for monitoring pulsatile signals such as those occurring in pulse oximetry (e.g., West,

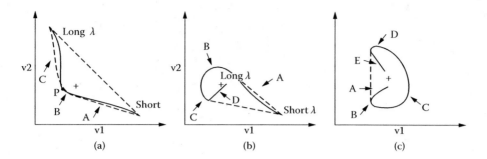

FIGURE 3.5
Chromatic diagrams x:y for processors with different response profiles: (a) $x = R_1/(R_1 + R_2 + R_3)$; $y = R_2/(R_1 + R_2 + R_3)$; (b) $x = R_1/(R_1 + R_3 + R_4)$; $y = R_3/(R_1 + R_3 + R_4)$; (c) $x = R_3/(R_3 + R_4 + R_5)$; $y = R_4/(R_3 + R_4 + R_5)$. (From Jones, G.R. and Russell, P.C., *Pure Appl. Opt.*, 2, 87–110, 1993. With permission.)

1993; Chapter 8, Section 8.2). For such cases the Cartesian coordinates may be chosen to be

$$R_1 = (r_0/b_0)/(B_0/R_0) \qquad (3.20)$$

$$R_2 = (g_0/b_0)/(B_0/G_0) \qquad (3.21)$$

where r_0, g_0, b_0 are the time varying outputs from each of the three processors and R_0, G_0, B_0 are the steady background outputs.

(e) Other Particular Forms of Tristimulus Algorithms

There are several other special cases of the algorithms represented by Equations 3.11 and 3.12 depending upon the values chosen for the weighting factors. Two examples that have been used for chromatic monitoring are given in the following text.

(i) Different Gains for Each Denominator Term

This case corresponds to the following conditions:

$$n_1 \neq n_2 \neq n_3$$

$$m_1 = m_2 = m_3 = 0$$

$$A = B = a = b = 1$$

Insertion of these values into Equations 3.11–3.12 lead to the following equations:

$$x = R_0/(n_1 R_0 + n_2 G_0 + n_3 B_0) \qquad (3.22)$$

$$y = G_0/(n_1 R_0 + n_2 G_0 + n_3 B_0) \qquad (3.23)$$

Such algorithms have been employed for optical fiber based chromatic monitoring of electrical plasma (e.g., Djakov et al., 2004; Khandaker et al., 1994). The effect of n_1, n_2, n_3 being non-zero is to distort the boundary containing the chromatic space (e.g., CIE boundary Figure 3.2; Bevan, 1989) and to shift the regions of maximum sensitivity to regions of the optical spectrum where priority information may reside.

(ii)   Different Gains for Denominator and Numerator
This case corresponds to the following conditions:

$$n_1 = n_2 = n_3 = 1$$

$$m_1 = m_2 = m_3 = m$$

Insertion into Equations 3.11–3.12 leads to Equations 3.24–3.25.

$$x = \frac{A[a\,R_0 - m(R_0 + G_0 + B_0)]}{(R_0 + G_0 + B_0)} \qquad (3.24)$$

$$y = \frac{B[b\,G_0 - m(R_0 + G_0 + B_0)]}{(R_0 + G_0 + B_0)} \qquad (3.25)$$

where $R_0 + G_0 + B_0 = L$.

These chromatic algorithms have been employed by Jones et al. (1994) for broadband interferometric monitoring with optical fibers (Chapter 6, Section 6.2.2.1).

The forms of particular algorithms represented by Equations 3.11 through 3.25 are summarized in Table 3.1.

TABLE 3.1

Values of Weighting Factors Equations 3.11–3.12, Corresponding to Each Algorithm of Section 3.3.2

System	Equations	A	B	a	b	m_1	m_2	m_3	n_1	n_2	n_3
(a) Distimulus	(3.13, 3.14)	1	1	1	1	0	0	0	1	1	0
(b) Tuned distimulus	(3.16, 3.17)	A	B	a	b	m_1	m_2	0	n_1	n_2	0
(c) Tristimulus CIE	(3.4, 3.5)	1	1	1	1	0	0	0	1	1	1
(e) (i) Tuned tristimulus I	(3.22, 3.23)	1	1	1	1	0	0	0	n_1	n_2	n_3
(e) (ii)Tuned tristimulus II	(3.24, 3.25)	A	B	a	b	M	m	m	1	1	1
(d) Modified coordinates	(3.20,3.21)	1	1	1	1	0	0	0	1	1	1

3.4 *Lab* Transformation

The *Lab* model was introduced in color science as an alternative perceptive approach to the *x, y, z* model to better reflect the operation of color processing by the human vision system. However, it has a number of attributes that can be usefully deployed for processing chromatic monitoring signals under some circumstances.

The *Lab* algorithms are defined by Equations 3.26–3.28 (Schwarz et al., 1987; Aïnouz et al., 2006):

$$L = 116 \ (G/Gn)*1/3 - 16 \tag{3.26}$$

$$a = 500 \ [(R/R_n)*1/3 - (G/G_n)*1/3] \tag{3.27}$$

$$b = 200 \ [(G/G_n)*1/3 - (B/B_n)*1/3] \tag{3.28}$$

$$(\text{for } R/R_n, \ G/G_n, \ B/B_n > 0.008856)$$

where R, G, B are outputs of each of the three processors (see Appendix 1.I). R_n, G_n, B_n are the processor outputs when addressing the illumination source directly.

These equations lead to a spherical chromatic space as shown in Figure 3.6. The L range (vertical axis, Figure 3.6) is from 0 (black) to 100 (white), the a axis

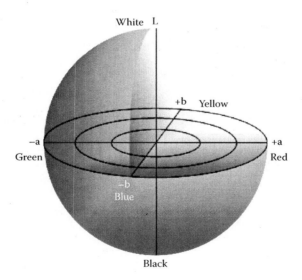

FIGURE 3.6
CIE *Lab* color space. (From Aïnouz, S., Zallet, J., de Martino, A., and Collet, C., *Opt. Express* 14(13), 5916–5927, 2006. With permission.)

extends from green ($-a$) to red ($+a$) and the b axis from blue ($-b$) to yellow ($+b$). Effectively, H and S values may be defined by Equations 3.29–3.30:

$$H_{ab} = \tan^*(-1)\,(b/a) \tag{3.29}$$

$$S_{ab} = (a^{*2}+ b^{*2})^{*1/2} \tag{3.30}$$

From the perspective of chromatic monitoring, the *Lab* system offers the following advantages over the x, y, z system:

(a) Achromatic point ($S = 0$) is at the origin (cf. Figure 3.2a).
(b) Chromatic space is more symmetrical (cf. Figure 3.2a).
(c) Form of the chromatic boundary does not vary with the character-istics of the processors (cf. Figure 3.5).

However, these features are offset by the *Lab* algorithms being mathemati-cally more complicated than the x, y, z algorithms which can be a disad-vantage for rapid, online monitoring in real applications, but less so for retrospective information extraction.

A further aspect of the *Lab* model is the form of signal normalization employed. With the basic x, y, z algorithms (Equations 3.4–3.6), the output of each processor (R, G, B) is normalized with respect to the overall signal strength ($R + G + B$) at each instant. With the *Lab* system, each processor output is normalized with respect to its own value from an a priori speci-fied condition (in color science, the source illumination). For chromatic monitoring this can provide additional discrimination capabilities as it has the effect of making more equitable the outputs from the three processors when the individual processor outputs are of substantially different magni-tudes. This can be significant when the change in the monitored condition is more focused within the response range of the processor with the lowest output. In some respects, the setting of different gains for each term in the Generalized x, y, z Chromatic Algorithms (Section 3.3.2, Equations 3.11–3.12) might be regarded as synonymous with the *Lab* normalization.

The *Lab* algorithms have been employed for tracking variations in the cen-ter of gravity location in simulations of flying aircraft using outputs from a number of different parameter sensors (Chapter 5, Section 5.4) deployed in a discrete chromatic processing manner (Chapter 2, Section 2.5).

3.5 Summary

Alternative algorithms for chromatic monitoring are based upon the H, S, V; x, y, z, and *Lab* schemes of color science.

The H, S, V algorithms are useful for discriminating changes when the bandwidth (saturation) variations dominate.

The x, y, z algorithms are convenient to implement in hardware or software for discriminating changes in monitored conditions.

They enable the result of combining two signals to be conveniently determined.

The x, y, z algorithms may be generalized by scaling each term relative to each other. (This effectively varies the relative disposition of each processor response with respect to each other (Chapter 2, Section 2.2)).

Consequently, the resolution of two or three processor systems can be made to emphasize different regions of a parameter domain.

With the x, y algorithms, changing the processor responsivities with respect to each other varies the shape of the monochromatic boundary whereas with the H, L, S and *Lab* algorithms the monochromatic boundary remains fixed as a circle.

The *Lab* algorithms provide a more symmetrical chromatic space than the x, y, z algorithms and a monochromatic boundary that does not vary with processor characteristics.

The *Lab* algorithms also normalizes the outputs from each individual processor separately, rather than collectively, and with respect to an a priori determined factor, rather than the instantaneous, overall signal strength $(R + G + B)$.

References

Aïnouz, S., Zallet, J., de Martino, A., and Collet, C. (2006). Physical interpretation of polarisation-encoded images by colour preview, *Opt. Express* 14(13), 5916–5927.

Bevan, C.M., 1989, Colour measurement in optical metrology, Ph.D. thesis, University of Liverpool.

Billmeyer, F.W. and Saltzman, M. (1981). *Principles of Color Technology*, John Wiley, New York.

Djakov, B., Harabovsky, M., Kopecky, V., and Jones, G. R. (2004). Monitoring water vapour plasma jets admixed with argon using chromatic sensing techniques, *High Temp. Mat. Proc.*, 8(2), 185–194.

Jones, G.R. and Russell, P.C. (1993). Chromatic modulation-based metrology, *Pure Appl. Opt.*, 2, 87–110.

Jones, G.R., Russell, P.C., and Khandakar, I.I. (1994). Chromatic interferometry for an intelligent plasma processing system, *Meas. Sci. Technol.*, 5, 639–647.

Jones, G.R., Li, G., Spencer, J.W., Aspey, R.A., and Kong, M.G. (1998). Faraday current sensing employing chromatic modulation, *Opt. Commn.*, 145, 203–212.

Jones, G.R., Russell, P.C., Vourdas, A., Cosgrave, J., Stergioulas, L., and Haber, R. (2000). The Gabor transform basis of chromatic monitoring, *Meas. Sci. Technol.*, 11, 489–498.

Khandaker, I.I., Glavas, E., Morse, K., Moruzzi, S., and Jones, G.R. (1994). Chromatic modulation as an online plasma monitoring technique, *Vacuum*, 45, 109–113.

Li, G.D., Aspey, R.A., and Jones, G.R. (1999). White LED based Faraday current sensor using a quartz wavelength encoder, *Opt. Commn.*, 162, 44–48.

Moon, P. and Spencer, D.E. (1981). *The Photic Field,* MIT Press, Cambridge, MA.

Murphy, M.M. and Jones, G.R. (1993). An intrinsic integrated optical fibre strain sensor, *Pure Appl. Opt.*, 2, 33–49.

Pavlova, P., Cyrrilov, K., and Moumdjiev, I. (1996). Application of HSV colour system in identification by colour of biological objects on the basis of microscopic images, *Medical Imaging and Graphics*, 20, 357–364.

Rogers, D. (1985). *Procedural Elements for Computer Graphic*, McGraw-Hill, New York.

Schwarz, M.W., Cowan, W.B., and Beatty, J.C. (1987). An experimental comparison of RGB, YIQ, LAB, HSV and opponent colour models, *ACM Transactions on Graphics*, 6, 552.

West, I. P. (1993). Optical Fibre Based Pulse Oximetry, Ph.D. thesis, University of Liverpool.

Section B

Deployment Examples

4

Chromatic Monitoring of Electrical Plasmas

B. Djakov, G.R. Jones, and Y. Yokomizu

CONTENTS

4.1 Introduction

Electrical plasmas are produced when gases become ionized, as a result of which they are electrically conducting and optically emitting. Examples of such plasmas, which occur naturally, are lightning strikes and the sun. Plasmas can also be produced for, or occur in, many areas of technology. Figure 4.1.1 shows images of some plasma forms that occur in various areas of

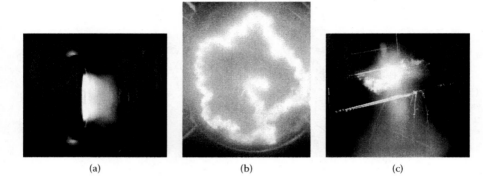

(a)   (b)   (c)

FIGURE 4.1.1
(See color insert following page 18). Examples of technological plasmas: (a) plasma used for processing materials (Courtesy Professor J. Bradley); (b) convoluted electric arc plasma in a rotary arc circuit breaker; (c) electric arc plasma on a high-voltage railway power line. (From Jones, G.R., Spencer, J.W., and Yan, J., in *Advances in High Voltage Engineering*, IEE, London, 2004, pp. 545–590. With permission.)

technology: Figure 4.1.1a shows a plasma used for processing materials; a convoluted electric arc plasma column occurring in electromagnetic, rotary arc circuit breakers; and an electric arc plasma occurring between the pantograph of a railway locomotive and an overhead power line.

Such plasmas are of a complex nature. For example, with electric arc plasmas (e.g., Figure 4.1.1b,c), the intensity of the radiation emitted varies over many orders of magnitudes; the composition involves many chemical components that interact in a complicated manner; the plasma boundaries are governed by the transfer of energy, mass, as well as momentum, and are of a complex geometry and affected by turbulence and various fluctuations (e.g., Jones and Fang, 1980). The manner in which such plasmas evolve is sensitive to small perturbations and sudden, abrupt changes can occur in the behavior pattern under similar operating conditions. Consequently, it is difficult to predict the behavior of such plasmas and also to adequately control them. It may be argued that this is one aspect of the problem that has inhibited the technological realization of plasma fusion for energy production.

The application of detailed reductionist methods (both theoretical and experimental) has elucidated many of the detailed properties and behavior of such plasmas (e.g., Jones and Fang, 1980). However, the problems of determining the global properties and behavior of such plasmas and in identifying the onset of performance-threatening perturbations in a cost-effective and robust manner for industrial use remains a challenge that can be addressed with chromatic techniques.

Chromatic methods can be deployed in either wavelength or space domains for monitoring technological plasmas. Wavelength-domain examples include the monitoring of low-pressure plasmas such as those used for processing semiconductor materials and for monitoring atmospheric pressure plasma jets within which particulates are entrained for modifying the properties of the surfaces of materials. Examples of the use of space-domain chromaticity are for monitoring the location of axisymmetric, filamentary arc plasmas, for determining the shape and movement of convoluted electric arc plasmas, and for tracking changes in the geometry of the boundaries of plasma jets.

4.2 Wavelength-Domain Chromaticity

4.2.1 Plasma Processing of Semiconductor Materials

Electrical plasma systems are used for depositing and etching thin layers of materials in the manufacture of semiconductor devices. The final product depends critically upon the plasma being in a well-defined state. This can be affected by small variations in a number of input parameters and can determine the yield of the process output (Russell et al., 1994).

Figure 4.2.1 shows a schematic of the substantial number of inputs and outputs for such a plasma processing system that can affect the magnitude

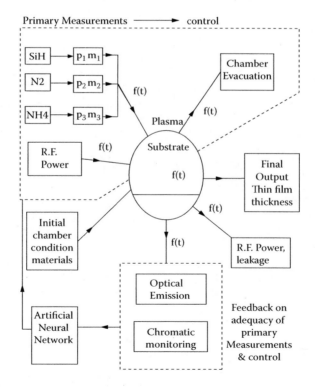

FIGURE 4.2.1

Schematic of a typical semiconductor plasma processing system showing monitoring and control functions. (From Russell, P.C., Alston, D., Smith, R.V., and Jones, G.R., *Nondestructive Testing and Evaluation*, 12, 1996. With permission.)

and the quality of the yield (Russell et al., 1996). These include the input (pressure p, mass flow m) of three gases (SiH, N_2, NH_4) in which the plasma is formed, the evacuation of the processing chamber, correct radio frequency (RF) power for activating the plasma, sufficiently clean materials, etc. Whereas each of these components can be measured separately outside the plasma chamber, there is nonetheless a need to ensure by direct monitoring that the plasma formed is indeed in the correct condition for maximizing the yield from the processing. For example, the level of RF power input may be directly measured, but the amount of power fed into the plasma rather than leaking around the chamber is uncertain. The standard method for the direct monitoring of the plasma is via the spectrum of the light it emits. A typical optical spectrum for such a plasma in hydrogen is given in Figure 4.2.2a (Russell et al., 1994). This illustrates the large number of data contained in such a spectrum. The structure of the spectrum also varies with operating conditions. Interpretation of the spectral pattern for controlling the system is therefore difficult. An alternative approach is to examine only one or two spectral lines. This provides a more rapid feedback for control purposes but may miss vital indications that may emerge in other parts of the spectrum. Monitoring the chromaticity of the plasma can overcome both these deficiencies in a conveniently deployable manner.

Figure 4.2.2b shows two optical spectra of a nitrogen plasma, one sustained by an RF power of 20 W and the other 40 W (Khandaker et al., 1994). The figure shows that differences between the spectra for the two powers are not located within narrow wavelength regions but are distributed over an extended wavelength range. Consequently, monitoring a couple of spectral lines does not yield a full indication of the scale of the change and does not necessarily provide a distinction between a change in the plasma power consumption and impurities present (an unknown component, e.g., hydrogen), contaminating a host plasma (e.g., nitrogen), each with a complex spectrum of the form shown in Figure 4.2.2a and 4.2.2b).

Although the use of full spectrum records may provide a visible distinction between the various conditions, they do not easily lend themselves to a simple quantification of the differences and would in any case involve much computational effort.

The application of chromatic monitoring provides a means for overcoming such difficulties. Addressing the optical emissions from the various plasmas with three nonorthogonal optical detectors enables three chromatic parameter values to be obtained, which classify the plasma in the chromatic domain. Figure 4.2.2c shows a chromatic map in terms of the $x{:}y$ chromatic parameters (see Chapter 3, Section 3.2) for a number of plasmas with different proportions of hydrogen and nitrogen, as well as for different RF powers in the range 20–40 W (Khandaker et al., 1994). Thus, each plasma condition has been identified by only two coordinates (x, y), which enables the different

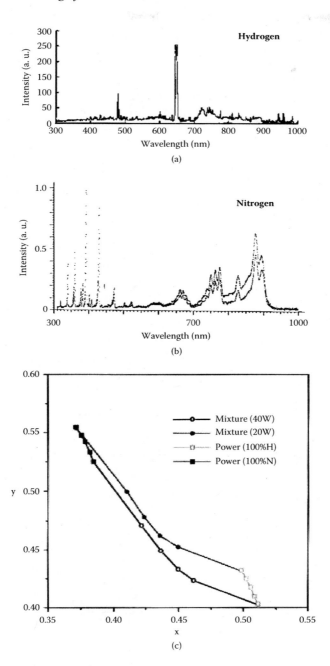

FIGURE 4.2.2

Optical spectra of processing plasmas and their x:y chromatic representation: (a) pure hydrogen plasma (Russell et al, 1994); (b) pure nitrogen plasma sustained by 20 W (lower trace) and 40W RF source (Khandaker et al., 1994); (c) chromaticity variation with different mixtures of H_2 and N_2 and with different RF plasma powers. (From Khandaker, I, Glavas, E., Morse, K., Moruzzi, S., and Jones, G.R., *Vacuum*, 45, 1, 1994. With permission.)

conditions to be conveniently quantified. The extent of the quantification and data compression is such that 18 spectra of the form shown in Figure 4.2.2a,b are easily displayed in the chromatic map of Figure 4.2.2c and distinguished from each other.

Two illustrations of the significance of this chromatic approach for plasma processing applications are given in Figure 4.2.3 and 4.2.4. Figure 4.2.3 shows an *x:y* chromatic map for a plasma used to etch a semiconductor sample (Khandaker, 1996). The locus of the points on this diagram represents the time evolution of the plasma chromaticity during the etching process. It clearly shows the stabilization of the etching plasma during which period impurities are desorbed from the sample to be etched. The onset and termination of etching are clearly discernible on the chromatic diagram. This illustrates the usefulness of the approach for conveniently tracking a plasma-etching process.

Figure 4.2.4 shows the variation of the three chromatic parameters, *H, L, S,* with RF power and gas pressure for a plasma enhanced chemical vapor deposition (PECVD) system (Russell et al., 1996). This illustrates the following trends:

- *H* decreases monotonically with both RF power and gas pressure.
- *L* increases with RF power and decreases with gas pressure.
- *S* decreases more steeply with RF power than with gas pressure.

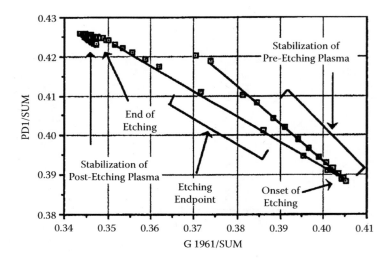

FIGURE 4.2.3
Variation in plasma chromaticity (*x:y* parameters) during the etching of a semiconductor substrate (Khandaker, 1996).

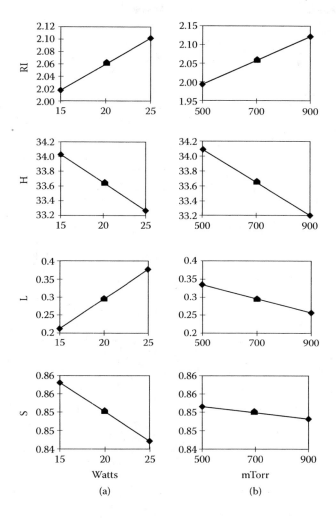

FIGURE 4.2.4
Variation in *H, L, S* chromatic coordinates and film refractive index (RI) with gas pressure and radio frequency power: (a) with RF power at constant pressure (500 mT); (b) with pressure at constant RF power (20 W). (From Russell, P.C., Alston, D., Smith, R.V., and Jones, G.R., *Nondestructive Testing and Evaluation*, 12, 1996. With permission.)

Also shown in Figure 4.2.4 is the variation of the refractive index (RI) of the deposited substrate, which increases with both RF power and gas pressure.

Since the variation of each chromatic parameter (*H, L, S*) are different functions of RF power and gas pressure, this provides a means for discriminating between drifts in these two latter parameters during a plasma-processing procedure. They, therefore, provide the basis for online control to bring the plasma to an optimum condition for yielding the required refractive index for the substrate.

4.2.2   Plasma Jets Monitoring

4.2.2.1   Introduction

Plasma spraying is a technique used to deposit on surfaces layers of materials, which are approximately 1–100 μm thick and have special properties (Gerdeman, 1972; Montavon, 2004). Examples are hard coatings for machine tools, ferromagnetic layers, and high-temperature superconductor ceramics.

A typical plasma-spraying system is shown in Figure 4.2.5. Powder particles are injected into a plasma flowing from a nozzle. They are entrained into the plasma, heated, melted, and then deposited onto the surface to form the coating.

The plasma in such torches is formed in a gas, such as air, argon, water vapor, etc. (Djakov et al., 2006a; Russell et al., 2000) flowing at rates of 5–100 slm, and maintained by electric currents of ~10 s to 100 s A.

The structure of such plasma jets is complex and variable with time (Figure 4.2.6a and 4.2.6b) not only because of the plasma chemical composition, the high temperatures (~2–30 × 10³ K), and the presence of particles, but also because of the flow being turbulent.

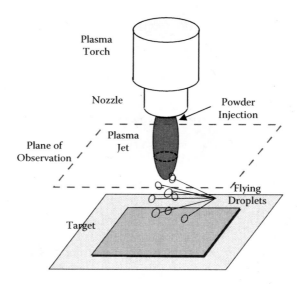

FIGURE 4.2.5
Schematic of a plasma-spraying system.

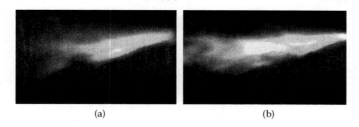

(a)   (b)

FIGURE 4.2.6
(See color insert following page 18). Sequential images of a plasma jet with particles (air—
75 slm, 200 A; exposure—0.5 ms; interframe time—several ms).

The physical properties of such plasma jets have been extensively studied using detailed spectroscopic techniques (Griem, 1964). Each spectrum corresponding to a single location and time instant contains typically 10^4 data points. To cover a number of radial and axial positions at several time intervals, at least 10^4 such measurements are needed leading to at least 10^8 spectral data points being captured. Whereas such an amount of data acquisition and processing may be justified for detailed, fundamental studies, it is uneconomic and impractical for online monitoring and control of industrial plasma processing.

The use of chromatic sensing and processing in the wavelength domain can provide the means for overcoming such difficulties of monitoring plasma conditions under robust industrial conditions.

4.2.2.2 Chromatic Processing Protocols

The optical monitoring of plasma jets (Djakov et al., 2004a,b, 2006a; Russell et al., 2000, 2002, 2003a,b) involves operating sensors in appropriate ranges of the electromagnetic spectrum and adopting suitable chromatic processor characteristics (see Chapter 2, Section 2.1).

Both visible and near-infrared portions of the electromagnetic spectrum may be addressed. The visible range (380–740 nm wavelengths) can be addressed using a CCD camera (Djakov et al., 2004b) with the R, G, B processor ranges defined as in Table 4.1 (VCS).

The near-infrared chromatic sensor (NCS) range (250–1700 nm), which also covers the visible and near-ultraviolet ranges, is addressed by three photodiodes R, G, B each covering the wavelength ranges shown in Table 4.1 (NCS). The NCS captured radiation from a 1×1 mm field of view at a chosen axial distance from the jet nozzle (plane of observation, Figure 4.2.5; Russell et al., 2002).

The characteristics of the chromatic processors (see Chapter 2, Section 2.1) can be adjusted to a limited extent by varying their relative amplitudes either in hardware (amplifier gain) or software. Figure 4.2.7a shows NCS results in the form of an H-S polar diagram for a 260 A arc plasma with an air flow

TABLE 4.1

Chromatic Processor Ranges and Characteristics

Sensor	Wavelength Range of R, G, B Channels, nm			Abbreviated (CS for Chromatic Sensor)
	R	G	B	
Photodiodes	700–1700	500–1200	250–950	NCS (near-infrared)
640 × 480 pixel CCD camera (Imaging source DFK21F04)	575–740	470–620	380–500	VCS (visible)

of 83 slm at an axial position of 18 mm from the jet nozzle. About 100 data points were gathered with a time resolution less than 1 ms and a repetition rate of approximately one per second. The variability of the plasma condition leads to an almost circularly symmetric distribution of points as shown in Figure 4.2.7a defined by $0 < H < 360$, $0.1 < S < 0.6$. One means for providing more discrimination in the H range is to vary the relative magnitudes of the R, G, B sensor responses. Figure 4.2.7b shows the effect of increasing the G and B sensor gains by a factor of 5 relative to the R signal. The signature of the plasma jet is then defined by a narrower H range of $130 < H < 220$ with S in the more monochromatic range of $0.5 < S < 0.9$.

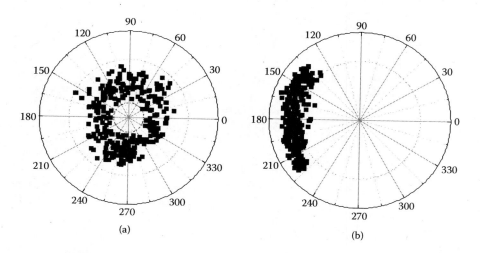

(a)   (b)

FIGURE 4.2.7

H-S polar diagrams for the same spectral conditions but different sensor gains (plasma current—260 A; air flow rate—83 slm; monitoring position—18 mm, on axis from nozzle exit): (a) R, G, B; (b) R, $5G$, $5B$. (From Djakov, B.E., Enikov, R., Oliver, D.H., Hrabovsky, M., and Kopecky, *Plasma Proc. Polym.*, 3, 2, 2006a. With permission.)

4.2.2.3   Nature of Chromatically Monitored Jet Spectra

Examples of optical spectra produced by a plasma torch (Russell et al., 2000) are shown in Figure 4.2.8a,b. These are for a plasma jet in air flowing at a rate of 16 slm, sustained by a current of 180 A and without powder injection. The two spectra correspond to two axial positions (27, 56 mm) from the nozzle. The distributed nature and complex nature of the spectra are apparent.

Applying the VCS sensing (Table 4.1) to the plasma conditions represented by these two spectra produces chromatic H, S coordinate values that represent each of the two conditions. These values are summarized in Table 4.2.

These results show that the signal spread in each case is similar ($S = 0.49$, 0.43) but that the dominant wavelengths are different ($H = 22°$, $78°$) enabling the two spectra to be distinguished.

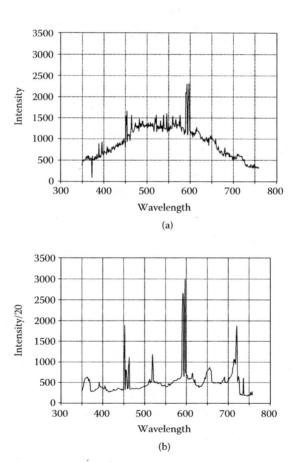

(a)

(b)

FIGURE 4.2.8
Spectra of axial portions of a jet (180A, air 16 slm, without powder) at two axial distances from the nozzle exit (a) –56 mm (b) –27 mm.

TABLE 4.2

Chromatic *H,S* Parameter Values Representing
the Conditions Corresponding to the Spectra
of Figures 4.2.8a and 4.2.8b

Position (mm)	*H* (degrees)	*S*
27	22	0.49
56	78	0.43

Application of processors with Gaussian responses (see Chapter 2, Section 2.2) that approximate the VCS sensor responses to theoretical spectra for air at temperatures from 2 to $20 \times 10^{3\circ}$K (Kamenshchikov et al., 1971) shows that the chromatic *H* and *S* parameters vary with temperature (Russell et al., 2000). Consequently, uniform bulk temperatures in this range could be deduced from *H* and *S* values.

In practice, experimental results show that for plasma jets, the *H, S* chromatic coordinates vary with the gas flow rate, plasma sustaining current, and gas type (e.g., argon, air), as well as the axial distance from the nozzle (Table 4.2).

Chromatic response surface diagrams may be obtained (Djakov et al., 2004b, 2006a) for distinguishing the operating condition of the plasma torch.

4.2.2.4   *H–S Polar Diagram of Plasma Torch Operation*

The various aspects of plasma jets, which need to be addressed to ensure that there is correct operation of the torch, include the dynamic signatures of the host plasma of the jet, the particles injected into the plasma, and material evaporated from the plasma torch electrodes. The plasma signatures can indicate deviations in plasma current, gas flow, etc. (Djakov et al., 2004a). The particle signatures can indicate whether the particles are entrained and adequately heated. The electrode vapor signature can indicate whether there is excessive electrode wear. In addition, the magnitude of fluctuations in the chromatic parameters of the plasma is an indication of imminent torch failure (Russell et al., 2000).

Chromatic signatures relating to the host plasma, particles, and electrode vapors are shown in the *H-S* polar diagram (Figure 4.2.9a) for an air plasma jet monitored with a properly tuned NCS chromatic sensor (Table 4.1). This figure shows that the three sources of chromatic signatures—pure plasma, heated particles, ionized electrode vapor—form three distinct groups on an *H-S* polar diagram. The signatures of each of these three components have been determined separately—the plasma signature by operation of the jet without particles; the particle signature both theoretically and experimentally for a "black-body" radiator

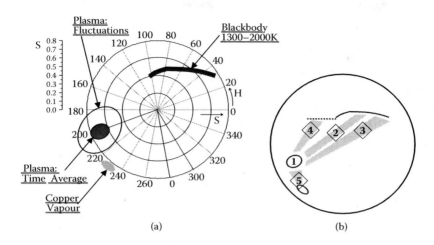

(a)   (b)

FIGURE 4.2.9
H-S polar diagram for a plasma spraying system operating with air: (a) chromatic regions of
the three components; (b) schematic of the superposition of the three signatures.

(Russell et al., 2000); the electrode vapor signature from copper vapor
spectra (Jones, 1983).

The pure plasma signatures lie in a range defined by $H \sim 200$ with fluctuations of $180 < H < 220$, $0.4 < S < 1.0$ (Figure 4.2.9a). The "Black Body" radiation from the particles lies in the range $30 < H < 100$, $0.4 < S < 0.8$, depending
on their temperature. The copper vapor signature is in a narrow band with
$H \sim 230$, $S \sim 0.9$.

In practice, each of these sources exists simultaneously so that the resultant chromatic signature is a combination of all three regions shown in
Figure 4.2.9a suitably weighted with respect to each other. The resultant
chromatic signature is related to the three individual signatures according
to the rules of chromatic superposition (see Chapter 3, Figure 3.2b). This
leads to the identification of the operating regime of a plasma torch from the
H-S coordinate range of its chromatic signature as shown schematically in
Figure 4.2.9b. Operation in each of the five regions shown in Figure 4.2.9b
has the following implications regarding control actions:

- Points 1 and 2 with small deviations—no action
- Point 1 approaches boundary of circle and/or is at 3 or 4—corrective action
- Point 5 and/or excessive jet fluctuations (outer circle for 1)—end
  operation

4.3 Space-Domain Chromaticity

4.3.1 Introduction

Monitoring the location and shape of electric arc filaments in various arc devices is important for comprehending performance trends of such units. The filamentary arcs are usually located within sealed enclosures of various types, which can have complicated structures (e.g., Jones, 2001a).

High-speed video cameras or optical multichannel analyzers have been utilized to measure the arc locations, shapes, and movement within such devices (e.g., Jones et al., 2004). However, these instruments have the disadvantages of being expensive and cumbersome, not providing a continuous time record of events and needing very large memories to record signals with high-time resolution and over a prolonged period. In addition, optical access is limited because the structure and material of the enclosure can play major roles in the arc extinction process and should not therefore be prejudiced to provide optical access.

Photodiodes or optical fibers can provide one means for limiting intrusive access provided the number of detectors/fibers employed can be minimized (Jones, 2001b). By using space-based chromaticity (see Chapter 2, Section 2.4) the number of detectors, which need to be deployed, can be optimized so as to reduce the amount of intrusion. In practice, the number of photodetectors may be limited to only three or four photodiodes, each addressing the interior of an enclosure from specifically chosen directions.

4.3.2 Axisymmetric Arc Filaments

There are electric arc devices, such as high-voltage circuit breakers, in which ideally the arc filament should be located along the axis of a nozzle for it to be controlled and extinguished by a flow of gas through the nozzle (Jones, 2001a). An arc located off the nozzle axis may cause residual high-temperature gas to exist off the central axis after arc extinction when the electric current has reduced to zero and can affect the current interruption performance of the circuit breaker. Conversely, moulded case circuit breakers, used in low-voltage distribution systems, drive a mainly linear filamentary arc from an ignition to a quenching region to limit the short-circuit current effectively. This operation causes the arc location to vary with time during the current interruption process. Space-domain chromaticity may be used for monitoring the location of such filamentary, electric arcs.

4.3.2.1 *Chromatic Methodology*

Figure 4.3.1a shows a side view of an axisymmetric arc filament between two electrodes within the nozzle of a gas flow circuit breaker. It also illustrates how optical access can be obtained to observe the arc filament section at

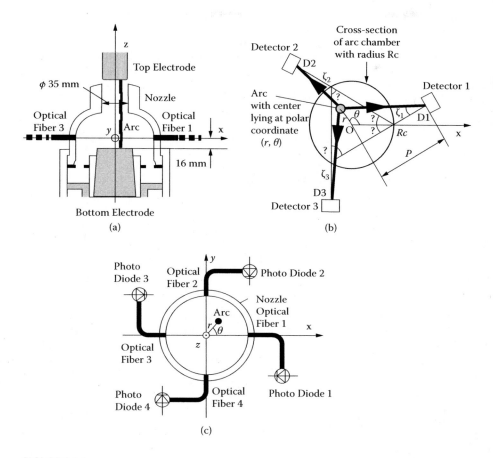

FIGURE 4.3.1
Various optical detectors deployments for filament arc monitoring: (a) cross-section *y:z* of inter-rupter nozzle showing axisymmetric arc and optical; fibers; (b) three optical detectors arranged in delta formation in the *x:y* plane; (c) four optical detectors in quadrative formation (From Yokomizu, Y., Spencer, J.W., and Jones, G.R., *J. Phys. D: Appl. Phys.*, 31, 1998. With permission.)

one axial (z) location. Figure 4.3.1b shows the deployment of three optical detectors in the *x-y* plane arranged in a "delta" formation (see Chapter 2, Figure 2.4.3). The arc filament of radius R_a is aligned orthogonal to the plane *x-y* along the *z*-axis (Figure 4.3.1a) and is located at a position with polar coordinates (r, θ) in the *x-y* plane. The optical emissions from the arc form optical signals for each of the three photodiode detectors, which have an identical polar sensitivity characteristic corresponding to one of the forms shown in Figure 4.3.2.

The output (I_j) from each of the photodiode detectors, j, $(j = 1,2,3)$ is given by (Yokomizu et al., 1998)

$$I(r^*, \theta, R_a^*) = k(S_w)Q_j(r^*, \theta, R_a^*, S_p) \qquad (4.3.1)$$

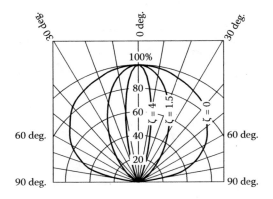

FIGURE 4.3.2
The relative polar sensitivity (Sp) of a photodiode as a function of the incident angle of light for various values of the parameter ζ. (From Yokomizu, Y., Spencer, J.W., and Jones, G.R., *J. Phys. D: Appl. Phys.*, 31, 1998. With permission.)

where k is a factor quantifying the response of the photodetector (including its wavelength responsivity, (S_w), $Q_j(r^*, \theta, R_a^*, S_p)$ is a factor incorporating the geometry of the device and collecting optics, r_θ^*) are normalized laboratory frame polar coordinates (Figure 4.3.1b) and R_a^* is the normalized arc radius and S_p is the polar sensitivity (Appendix 4.I).

The outputs (Equation 4.3.1) from the three photodetectors, I_1, I_2, I_3 (Figure 4.3.1b) are combined algorithmically to yield three chromatic parameters H, S, L (see Chapter 1), which in the present case of space chromaticity (see Chapter 2, Section 2.4) have the following meaning:

$H \rightarrow (\theta_d)$ is the dominant azimuthal angle (degrees).

$S \rightarrow (r_n)$ is the nominal radial coordinate.

$L \rightarrow (I_1 + I_2 + I_3)/3$ is a measure of the strength of the emission from the arc.

For position determination, it is θ_d, r_n that are the relevant chromatic parameters. By analogy with the definitions of H and S (see Chapter 1) expressions for θ_d and r_n are as follows:
when I_1, namely Q_1, is the lowest,

$$\theta_d(r^*, \theta, R_a^*) = 240 - 120 \frac{I_2 - I_1}{(I_2 - I_1) + (I_3 - I_1)} = 240 - 120 \frac{Q_2 - Q_1}{(Q_2 - Q_1) + (Q_3 - Q_1)}$$

(4.3.2)

when I_2, namely Q_2, is the lowest,

$$\theta_d(r^*, \theta, R_a^*) = 360 - 120 \frac{Q_3 - Q_2}{(Q_3 - Q_2) + (Q_1 - Q_2)}$$

(4.3.3)

when I_3, namely Q_3, is the lowest

$$\theta_d(r^*, \theta, R_a^*) = 120 - 120 \frac{I_1 - I_3}{(I_1 - I_3) + (I_2 - I_3)} = 120 - 120 \frac{Q_1 - Q_3}{(Q_1 - Q_3) + (Q_2 - Q_3)}$$

(4.3.4)

The nominal radius r_n is given by

$$r_n(r^*, \theta, R_a^*) = \frac{\max(I_1, I_2, I_3) - \min(I_1, I_2, I_3)}{\max(I_1, I_2, I_3) + \min(I_1, I_2, I_3)} = \frac{\max(Q_1, Q_2, Q_3) - \min(Q_1, Q_2, Q_3)}{\max(Q_1, Q_2, Q_3) + \min(Q_1, Q_2, Q_3)}$$

(4.3.5)

where θ_d is an angle increasing in the counterclockwise direction from zero along the x-axis (Figure 4.3.1b) and r_n increases from zero at the center of the enclosure to unity at the enclosure boundary.

Figure 4.3.3a shows theoretically determined characteristic curves of θ_d and r_n on a chromatic polar diagram for the delta formation of detectors shown in Figure 4.3.1b and for detectors with response lobes corresponding to $\zeta = 0$ (Figure 4.3.2) and for $R_a^* < 0.1$. The laboratory frame coordinates of a filamentary arc may be determined by locating the θ_d and r_n coordinates of the arc on the characteristic curves of Figure 4.3.3a and reading the values of θ and r^* from the calculated curves. For example, for values of the delta formation detector outputs I_1, I_2 and I_3 of 3.15, 2.16 and 1.97, the dominant angle θ_d and the nominal radius r_n are calculated to be 16.3° and 0.231, respectively (P in Figure 4.3.3a). These lead to the r^* and θ coordinates for the center of the arc cross section of 0.27 and 0.393 radian, respectively.

4.3.2.2   Verification and Parameter Effects

The validity of the space-domain chromatic approach for monitoring linear filamentary arcs has been verified experimentally and the effect of changes in operational parameters such as the radius of the arc filament and the polar sensitivity and geometric deployment of the photodetectors have been explored theoretically.

(a)   Experimental Verification

The approach has been verified by addressing a 1 mm diameter, glowing, tungsten filament (to replicate an arc) with delta deployed optical fibers and detectors.

The dominant angle and the nominal radius determined from the optical signals yielded the polar coordinates (r, θ) of the filament location as described in Section 4.3.2.1. Typical results are shown in Figure 4.3.4a for the r, θ values so determined as a function of the directly measured r coordinate and for the particular case of $\theta = \pi/2$, r variable. The good agreement found

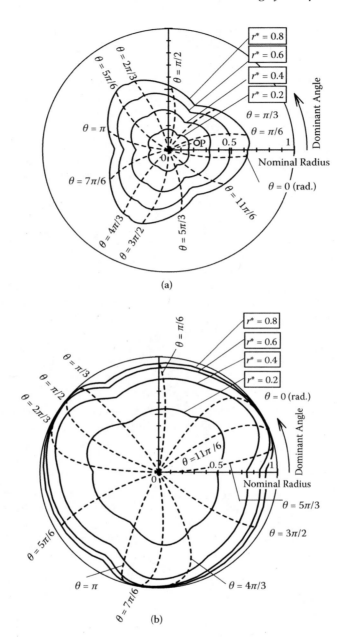

FIGURE 4.3.3
Chromatic maps (θ_d, r_n) for determining the position of the arc axis in the laboratory frame (r^*, θ). $R_a^* = 0.1$ and $p^* = 2$ in all cases. Full lines—curves of constant r^*; broken lines—curves constant θ. Detectors in delta formation (Figure 4.3.1b): (a) $\zeta = 0$; (b) $\zeta = 4$. (From Yokomizu, Y., Spencer, J.W., and Jones, G.R., *J. Phys. D: Appl. Phys.*, 31, 1998. With permission.)

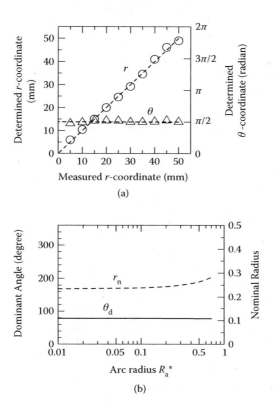

FIGURE 4.3.4
Verification and effect of parameters: (a) calibration test results for a tungsten filament with the delta configuration of detectors (Figure 4.3.1b); filament located at $\theta = \pi/2$. r, θ: radius and azimuthal coordinates in the laboratory frame of reference; (b) the effect of normalized arc radius R_a on the dominant angle (θ_d) and nominal radius (r_n) for the delta detector configuration (Figure 4.3.1b). (Laboratory polar coordinates of arc axis = (0.3, $\pi/2$); Detector polar sensitivity as shown in Figure 4.3.2 ($\beta = 0$). (From Yokomizu, Y., Spencer, J.W., and Jones, G.R., *J. Phys. D: Appl. Phys.*, 31, 1998. With permission.)

for both r and θ is replicated for cases when the lamp filament was placed at different r locations and with $\theta = 0$, π, $3\pi/2$ radians.

(b)   Polar Sensitvity of Detectors

The effect of varying the polar sensitivity of the detectors is synonymous with varying the responsivities of the detectors in the chromatic wavelength domain (see Chapter 2, Section 2.2). The effect of varying the polar sensitivity upon the dominant angle and the nominal radius may be examined theoretically using a general polar sensitivity $S'_p(\tau)$ expressed in terms of the basic sensitivity $S_p(\tau)$ and a variable term $\exp(-\zeta^2\tau^2)$ according to the following relationship:

$$S_p'(\tau) = \exp(-\zeta^2 \tau^2) S_p(\tau), \tag{4.3.7}$$

Figure 4.3.2 shows $S_p'(\tau)$ for various values of ζ in the range $\zeta = 0$–4.0. Figure 4.3.3b shows the dominant angle and the nominal radius for $\zeta = 4.0$ in the case of the delta formation (Figure 4.3.1b). Comparison of these characteristics with those for $\zeta = 0$ (Figure 4.3.3a) suggests that for this photodetector arrangement $\zeta \leq 0.4$ provides a reasonable range and sensitivity for locating an arc filament.

(c)   Arc Radius

The above analyses assumed that the arc filament had a radius of only 10% of the chamber/nozzle radius ($R_a^* = 0.1$). The effect of different normalized arc radii may be considered by calculating θ_d and r_n for an arc with its axis located at $r = 0.3$, $\theta = \pi/2$ and with normalized radii in the range $0.01 < R_a^* < 0.7$ for the delta detector arrangement (Figure 4.3.1b). The results of such calculations show (Figure 4.3.4b) that the dominant angle θ_d remains constant, regardless of the arc radius R_a^* whereas the nominal radius r_n is independent of R_a^* in the range 0.01–0.1, and increasing by 17% within the range 0.1–0.7. The implication of the result is that the characteristic curves shown in Figure 4.3.3a and calculated for $R_a^* = 0.1$ also apply for locating the axis of any arc with $R_a^* \leq 0.1$ with a high degree of accuracy.

(d)   Star Deployment of Detectors

The photodetectors may be deployed in an inverse star formation (i.e., aligned to monitor radially inwards; cf. Chapter 2, Section 2.4) rather than in delta formation. Optionally four rather than three detectors may be utilized to give a quadrature system (Figure 4.3.1c). The analytical form of the geometric factor $Q_j(r^*, \theta, R_a^*, S_p)$ is modified from that of the delta formation to take account of the different perspectives of the detectors (Appendix 4.I).

The quadrature deployment has been used for tracking the location of a time-varying arc filament within the nozzle of a high-voltage circuit breaker (Figure 4.3.1a) by processing the outputs from three of the quadrature photodiodes to yield the r_n, θ_d chromatic parameters (Yokomizu et al., 1998).

4.3.3   Convoluted Arc Columns

4.3.3.1   The Nature of Convoluted Arcs

Operation of rotary arc circuit breakers (e.g., Ryan and Jones, 1989) and arc heaters involves the formation of helically shaped arc filaments in an annular space between a rod electrode along the axis of a cylinder ($r = 0$) and the cylinder wall (radius R_c), which acts as the other electrode (Figure 4.3.5). The electric arc is rapidly rotated electromagnetically via the interaction of the current through the arc with a driving magnetic field. This interaction causes the location of the arc to continuously vary azimuthally and also results in

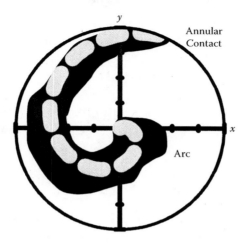

FIGURE 4.3.5

Frame from high-speed photograph of a convoluted arc column viewed from along the system axis and showing the system geometry. (From Yokomizu, Y., Jones, G.R., Spencer, J.W., and Yuan, W.D., *Proceedings of the 13th International Conference on Gas Discharges and Their Applications,* 1, 2000. With permission.)

the arc having various shapes, which, when observed in a direction along the central axis of the annular contact, may be convoluted, curved, or radial (Ryan and Jones, 1989). The complexity of these effects and their continuously varying nature make it difficult to define the arc and its behavior at an acceptable level of simplicity.

One example of the shape of a convoluted arc column is given in Figure 4.3.5, which shows a view along the axis of the cylindrical enclosure. One such convoluted form may be defined mathematically by the radial coordinate, r, being linearly related to the azimuthal coordinate, θ (Appendix 4.II). For convolutes, which are members of such a group, the shape of the convoluted arc filament is governed by a parameter, α, and its angular position by another parameter, β (Appendix 4.II).

Figure 4.3.6 shows a number of convoluted curves of different shapes and positions, all being members of the same $r = 0$ family but with different values of α and β. The shapes range from being purely radial to filaments, which overlap azimuthally.

4.3.3.2 Chromatic Polar Diagrams

Space-domain chromaticity may be utilized with the simplified description of convoluted arc columns afforded by the two convolution defining parameters α and β. The output from each photodetector monitoring a convoluted arc filament may be expressed as (see Appendix 4.I):

$$I_j(\alpha, \beta, R_a) = k(S_w)\, Q_j(\alpha, \beta, R_a, S_p) \tag{4.3.8}$$

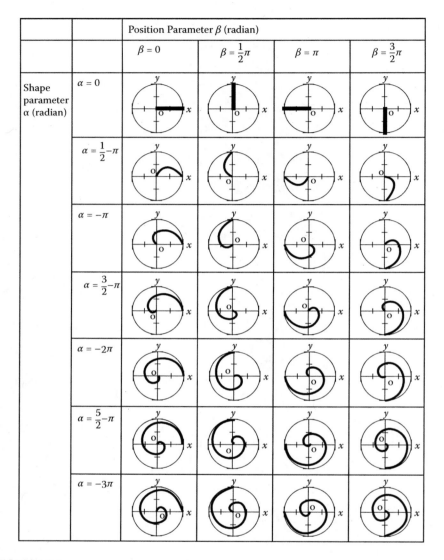

FIGURE 4.3.6
Classification of the shape and position of convoluted arcs: rotation—counterclockwise; arc—bold lines; annular contact—circle. (From Yokomizu, Y., Jones, G.R., Spencer, J.W., and Yuan, W.D., *Proceedings of the 13th International Conference on Gas Discharges and Their Applications*, 1, 2000. With permission.)

For three detectors with the same spectral sensitivities placed symmetrically at different azimuthal positions perpendicular to the central axis of an arc chamber and with their polar sensitivities nonorthogonal (Yokomizu et al., 2000), their outputs can be processed to yield two chromatic parameters—dominant angle, θ_d, and nominal radius, r_n—using equations of the same

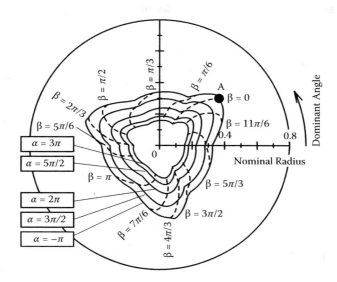

FIGURE 4.3.7
Chromatic map with coordinates r_n and θ_d: full lines—curves of constant α; broken lines—curves of constant β; $R_c = 1$; $R_a = 0.1$, $p = 2$. Polar sensitivity shown in Figure 4.3.2 ($\zeta = 0$). (From Yokomizu, Y., Jones, G.R., Spencer, J.W., and Yuan, W.D., *Proceedings of the 13th International Conference on Gas Discharges and their Applications,* 1, 2000. With permission.)

form as for the axisymmetric arc filament position determination (Equations 4.3.2–4.3.5).

Thus, convoluted arcs of different shapes (α) and positions (β) may be represented by different points (r_n, θ_d) on a chromatic polar diagram. An example of the characteristic curves, which emerge from theoretical calculations, is given on the chromatic polar diagram of Figure 4.3.7. For instance, an arc with $\alpha = -\pi$ has a dominant angle θ_d of 37.9° and a nominal radius r_n of 0.471 for $\beta = 0$ (shown as a filled circle 'A' in Figure 4.3.7). As this arc rotates, it follows the $\alpha = -\pi$ locus leading to r_n varying in the range 0.327–0.479 and forms a unique closed curve. The results of Figure 4.3.7 indicate that chromatic mapping of the outputs from three delta arranged photodetectors should enable both the shape and position of a convoluted, rotating arc to be determined via the convoluted parameters α, β.

4.3.4 Detailed Variations of Plasma Jet Boundary

A knowledge of the detailed lateral movement of the boundary of a plasma jet (Figure 4.2.5) provides information about the stability of the jet and hence the efficiency of its operation. Monitoring such detailed displacements can be accomplished with a particular form of space-domain chromaticity.

The displacement of the shaded ring (Figure 4.3.8a) representing the boundary of the plasma jet of circular cross-section is monitored with three

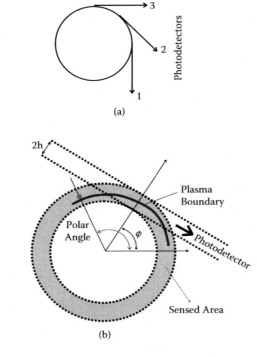

(a)

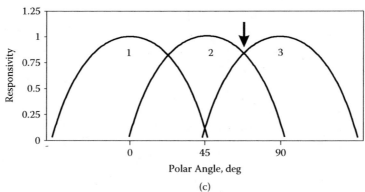

(b)

(c)

FIGURE 4.3.8
Deployment of space chromatic detectors for monitoring the position and movement of a plasma boundary: (a) relative inclinations of the three chromatic detectors; (b) schematic of the monitored boundary region; (c) angular responsivities of the chromatic detectors with respect to each other.

optical detectors inclined at 45° to each other (Figure 4.3.8b) while each
addresses the annular boundary tangentially (Djakov et al., 2004b,c; 2006).
Consequently, there are overlaps in the section of the annular ring from
which each detector receives signals. When the annular ring is displaced
or distorted, the signals received by the detectors assume different proportions, which depend upon the exact nature and magnitude of the perturbation. This mapping of such 2-D regions is similar to tomography (Ingesson
and Pickalov, 1996) but utilizes nonorthogonal rather than highly collimated
detectors to provide a simpler and more economic approach.

The responsivity of each detector with respect to the angle of inclination
from the first detector is as shown in Figure 4.3.8c. The movement or distor-
tion of the annular ring in the region of observation along the radius, ρ, of the
plasma disc and azimuthal angle, θ, may then be chromatically transformed
into H, S coordinates in a similar manner to that described in Section 4.3.2.1
(Equation 4.3.2–4.3.5) and displayed on an H-S polar diagram (Figure 4.3.9a).
Loci of constant radial displacement and azimuthal angle allow the observed
displacement to be determined in the laboratory frame coordinates, ρ, θ in a
similar manner to that described in Section 4.3.2.1. Figure 4.3.9b shows the
zones of highest, M_1, and lowest, M_2, resolutions for determining the laboratory frame coordinates ρ, θ. The highest resolution region corresponds to the

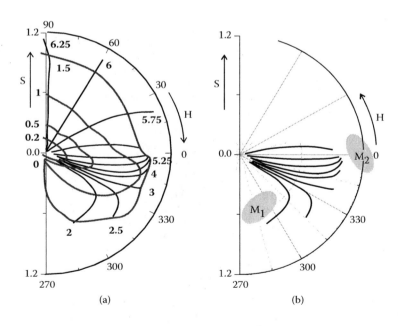

FIGURE 4.3.9
H-S polar diagram for converting chromatic H, S values into laboratory frame coordinates ρ, θ.
(a) curves of constant θ ($2 < \theta < 6.25$) are mainly radial; curves of constant ρ ($0 < \rho < 1.5$) are
mainly azimuthal; jet cross-section is a circle of unit radius (b) identification of zones of highest
(M_1) and lowest (M_2) resolutions on the H-S diagram showing also curves of constant θ.

point of maximum overlap of the detector responses, indicated by the arrow in Figure 4.3.8c.

4.4 Summary

Both optical wavelength and spatial domain chromaticity can be deployed for monitoring different types of electrical plasma utilized for various technological applications.

The deployment of optical wavelength chromaticity for monitoring low-pressure plasma for semiconductor wafer processing can identify departures from established operating conditions at sufficiently early stages to form the basis of online control of processes. The nature and magnitude of the chromatic change can indicate the basic cause of the change, distinguish between defective gas mixing, pressure drifts, platen temperature variations, leakage of RF power sustaining the plasma, etc.

In the case of plasma jets, optical wavelength chromaticity can distinguish changes distributed across substantial parts of the wavelength range or changes at specific wavelengths, which may not be obvious candidates for detailed spectroscopic measurements. Particulates heated by the plasma can be monitored and undesirable effects such as electrode evaporation can be distinguished.

Space chromatic methods can be used, not only for point source location (see Chapter 2, Section 2.4), but can also be deployed for locating axisymmetric filamentary arc plasma columns within the inaccessible interiors of enclosures such as gas flow controlling nozzles and for classifying the spiral shapes of convoluted arc filaments, which can occur in high voltage circuit breakers. The approach can also be deployed for detecting in more detail displacements of the boundary of a plasma jet.

Appendix 4.I   Detector Output Analysis

I.1   General Formulation

The output I_j from a space chromatic detector j may be expressed as the product of a signal response factor k and a system geometry factor Q_j.

$$I_j(f, g, R_a) = kQ_j (f, g, R_a, S_p) \qquad (I.1)$$

where f and g are the required arc position- or shape-defining parameters and R_a is the arc radius. The factor, k, is expressed in the following form (cf. Appendix 1.I):

$$k = U_F L_F \int A_F(\lambda) S_w(\lambda) \varepsilon(\lambda) \, d\lambda \qquad (I.2)$$

where U_F = cross-sectional area of the optical fiber core

L_F = length of the optical fiber

$A_F(\lambda)$ = attenuation by the optical fiber

$S_w(\lambda)$ = wavelength responsivity of the detector

$\varepsilon(\lambda)$ = optical emissivity of the arc

The factor, k, is independent of f and g. On the other hand, the factor, Q_j, is written as follows:

$$Q_j\,(f,\,g,\,R_a,\,S_p) = F_{lat}\,(G,\,S_p) + F_{cro}\,(G,\,S_p) \tag{I.3}$$

where G is a system geometry factor (Figures 4.3.1b and 4.3.1c) and S_p is a detector polar sensitivity factor (e.g., Figure 4.3.2).

I.2   Special Cases

I.2.1   Axisymmetric Filament, Delta Detector Deployment (Section 4.3.2.1)

Limited axial view $\longrightarrow F_{cro} \approx 0$

No optical fibers $\longrightarrow L_F = 1$ and $A_F(\lambda) = 1$ in Equation (I.2)

For position determination $f = r^*,\, g = \theta$

(r^*, θ are normalized radial and azimuthal coordinates, respectively)

$$Q_j\,(r,\,\theta,\,R_a,\,S_p) = F_{lat}\,(G,\,S_p) \tag{1.4}$$

where G and S_p are determined from the system geometry (Figure 4.3.1b) and the polar sensitivity of the detector (Figure 4.3.2). Analytical expressions for G and S_p are given by Yokomizu et al. (1998).

I.2.2   Axisymmetric Filament, Quadrature Detector Formation (Section 4.3.2.2)

With optical fibers $U_F,\, L_F,\, A_F(\lambda) < 1$

$$f = r^*,\, g = \theta$$

Detailed analytical expressions for $Q_j(r^*,\, \theta,\, R_a,\, S_p)$ (Equation I.3) are given by Yokomizu et al. (1998).

I.2.3   Convoluted Filament, Delta Detector Formation (Section 4.3.3)

No optical fibers $\rightarrow L_F = 1$ and $A_F(\lambda) = 1$ in Equation (I.2).

For position and shape determination $f = \alpha,\, g = \beta$ where α is the convolute shape factor, β is the convolute position factor.

Detailed analytical expressions for $Q_j(\alpha, \beta, R_a, S_p)$ (Equation I.3) are given by Yokomizu et al. (2000).

Appendix 4.II   Two-Parameter Description of a Convoluted Arc

One class of convolutes may be defined mathematically by

$$r = \theta \tag{II.1}$$

This equation may be transformed via a characteristic parameter ξ^* defined by

$$\xi^* = r^* = r/R_c \tag{II.2}$$

$$\xi^* = \theta^* = \theta/\alpha \qquad (\alpha \neq 0)$$
$$= 0 \qquad (\alpha = 0) \tag{II.3}$$

with $0 \leq \xi^* \leq 1$.

Convolute Shape

The parameter α defines the arc shape:

$\alpha = 0$ corresponds to a "radial" arc

$\alpha = 2\pi$ is for the convolute having one complete circumference within R_c

$\alpha = \pi$ yields a convolute having only half the circumference within R_c

(R_c = radius of the cylinder)

Negative α values correspond to the convolute curvature being oppositely configured.

Convolute Position

The angular position of the convolute may be described by a parameter β. where $\beta = \theta$ at $r/R_c = 1$.

The quantities α and β are related by the equation

$$\theta - \beta = \alpha(\xi^* - 1)$$

The various shapes and positions of convolutes corresponding to different values of α and β are shown in Figure 4.3.6.

References

Djakov, B.E., Hrabovsky, M., Kopecky, V., and Jones, G.R. (2004a). Monitoring water vapour plasma jets admixed with argon using chromatic sensing techniques, *High Temp Plasma Processes*, 8, 185–193.

Djakov, B.E., Enikov, R., Oliver, D.H., and Vasileva, E. (2004b). *In Annual Report of the Institute of Electronics*, Sofia.

Djakov, B.E, Oliver, D.H, Enikov, R, and Vasileva, E. (2004c). Optical monitoring geometrical properties of plasma jets by chromaticity methods, *XII Workshop on Plasma Technology* (Ilmenau, Germany), pp. 111–118.

Djakov, B.E., Enikov, R., Oliver, D.H., Hrabovsky, M., and Kopecky, V. (2006). Chromatic monitoring of DC plasma torches: The latest developments, *Plasma Proc. Polym.*, 3(2), 170–173.

Djakov, B.E., Oliver, D.H , Enikov, R., Vasileva, E. (2006). A simple optical monitoring technique for determining the geometrical characteristics of a plasma jet. *Plasma Proc. Polym.*, 3(2), 160–164.

Gerdeman, D. A. and Hecht, N. L. (1972). *Arc Plasma Technology in Materials Science*, Springer, Vienna, Austria.

Griem, H.R. (1964). *Plasma Spectroscopy*, McGraw-Hill, New York.

Ingesson, L.C. and Pickalov, V.V. (1996). An iterative projection-space reconstruction algorithm for tomography systems with irregular coverage, *J. Appl. Phys. D: Appl. Phys.* 29, 3009–3016.

Jones, G.R. (1983). High current arcs at high pressures, *16th International Conference on Phenomena in Ionised Gases*, Duesseldorf, Invited papers, pp. 106–118.

Jones, G.R. (2001a). CH 7 gas-filled interrupters—fundamentals, in *High Voltage Engineering and Testing*, Ryan, H.M., Ed., IEE, London, pp. 273–299.

Jones, G.R. (2001b). CH 21 optical fibre-based monitoring of high voltage power equipment, in *High Voltage Engineering and Testing*, Ryan, H.M., Ed., IEE, London, pp. 601–634.

Jones, G.R. and Fang, M.T.C. (1980). The physics of high power arcs, *Rep. Prog. Phys.*, 43, 1415–1465.

Jones, G.R., Spencer, J.W., and Yan, J. (2004). CH 12 optical measurements and monitoring in high voltage environments, in *Advances in High Voltage Engineering*, Haddad, M. and Warne, D., Eds., IEE, London, pp. 545–590.

Kamenshchikov, V.A., Plastinin, J.A., Nikolaev, V.M., and Novicky L.A. (1971). *Radiation Properties of Gases at High Temperatures*, Mashinostroenie, Moscow.

Khandaker, I. (1996). Optical fibre sensors for optimisation of plasma processing, Ph.D. thesis, University of Liverpool.

Khandaker, I, Glavas, E., and Morse, K., Moruzzi, S., and Jones, G.R. (1994). Online plasma monitoring, Vacuum, 45(1), 109–113.

Montavon, G. (2004). Recent developments in thermal spraying, *High Temp. Plasma Processes*, 8, 45–93.

Russell, P.C., Khandaker, I., Glavas, E., Alston, P., Smith, R.V., and Jones, G.R. (1994). Chromatic monitoring for the processing of materials with plasma, *IEE Proc. Sci. Meas. Tech.*, 141(2), 99–104.

Russell, P.C., Alston, D., Smith, R.V., and Jones, G.R. (1996). Online fibre optic-based inspection using chromatic modulation for semiconductor materials processing, *Nondestructive Testing and Evaluation*, 12, 379–389.

Russell, P.C., Djakov, B.E., Enikov, R., Oliver, D.H., and Jones, G.R. (2000). Chromatic monitoring of arc plasma spraying, *VIII Workshop on Plasma Technology* (Ilmenau, Germany), pp. 22–27.

Russell, P.C., Djakov, B.E., Enikov, R., Oliver, D., and Jones G.R. (2002). Chromatic monitoring of arc plasma spraying, *Proc. ICGD2002*, 2, 204–207.

Russell, P.C., Djakov, B.E., Enikov, R., Oliver, D., Wen, Y., Jones, G.R. (2003a). Monitoring plasma jets containing micro-particles with chromatic techniques, *Sensor Review* 23(1), 60–65.

Russell, P.C., Djakov, B.E., Enikov, R., Oliver, D.H., and Jones, G.R. (2003b). Chromatic monitoring of plasma jets used for plasma spraying, *Vacuum*, 69, 125–128.

Ryan, H.M. and Jones, G.R. (1989). SF_6 *Switchgear,* Peter Peregrinus, London, pp. 37–43.

Yokomizu, Y., Spencer, J.W., and Jones, G.R. (1998). Position location of a filamentary arc using a tristimulus chromatic technique. *J. Phys. D: Appl. Phys.* 31, 3373–3382.

Yokomizu, Y., Jones, G.R., Spencer, J.W., and Yuan, W.D. (2000). Optical fibre monitoring of convoluted arcs, *Proceedings of the 13th International Conference on Gas Discharges and Their Applications* (Glasgow), 1, 222–225.

5

Chromatic Monitoring of Industrial Liquids

J.W. Spencer

CONTENTS

5.1   Introduction

Chapter 4 described the deployment of chromatic techniques for monitoring electrical plasma based upon the optical emissions they produce. This chapter describes the use of chromatic methods for monitoring liquids whereby the spectral envelope of light from a controlled source is affected by a liquid through which the light propagates or by reflection from a liquid interface. Wavelength sensing of color changed chemical indicators, e.g., pH acidity indicators, is also described, as well as the use of discrete chromatic processing (see Chapter 2, Section 2.5) of signals from an array of liquid property sensors. The sensing is either by optical fiber, remote, or an array of conventional sensors and transducers.

5.2 Chromatic Optical Fiber Monitoring

Chromatic-based fiber optic sensors have been used for monitoring different liquids that include vacuum pump oil, various aviation fuels, immiscible layers, organic solvents, and the acidity of liquids. The sensor systems use optical fibers to transmit light to and from a sensor with the returned light being detected via a discrete number of photodetectors. The chromatic-based fiber optic sensors use multimode optical fibers (> 50 µm diameter) to enhance coupling efficiencies between the source, detector, and sensor. Such large diameter fibers also permit greater levels of optical power to be delivered to the optical sensor and to the detectors, thereby enhancing the signal-to-noise ratio of the optical system. Furthermore, the use of multimode fibers also affords some immunity to chromatic changes as a result of fiber bending and susceptibility to modal noise. For applications where higher light levels are required fiber bundles are also used (Jones et al., 2000).

5.2.1 Degradation of Vacuum Pump Oil

Monitoring the degradation of vacuum pump oil involves a broadband tungsten halogen source, fiber optic bundles and a distimulus chromatic detection unit (Khandaker, 1993). The spectral distribution of the light returning to the detectors is characterized by a dominant wavelength, λ_d (see Chapters 1 and 3), which is a function of the oil purity. Changes in the dominant wavelength are therefore indicative of the oil purity according to the relationship (Khandaker, 1993)

$$x = \text{constant} \ \frac{K_x A_x e^{-b_x(\lambda_d - \lambda_{p1})^2}}{K_x A_x e^{-b_x(\lambda_d - \lambda_{p1})^2} + K_y A_y e^{-b_y(\lambda_d - \lambda_{p2})^2}} \tag{5.1}$$

where K_x, K_y, A_x, A_y, b_x, b_y are constants of proportionality; and λ_{p1} and λ_{p2} are defined wavelengths relating the Gaussian responsivity of the two detectors R_x and G_y, respectively, of the distimulus system.

However, Equation 5.1 does not convey the true complexity of the absorption spectra in the various oils. The transmission spectra for pure and impure industrial pump oils are shown in Figure 5.2.1.1 (a,b,c) for three different oils (Edwards 17, Edwards 15, and Crylin 87), respectively. The differences in the optical absorption of the various oils are distributed throughout the wavelength range 400–1000 nm. The quantification of these spectra by the distimulus detection system via the dominant wavelength and the subsequent use of Equation 5.1 yields a chromatic parameter x. The values of this parameter for different oils and conditions are shown in Figure 5.2.1.2. There are differences between the three oils in their pure and impure states. For all three oils the chromatic parameter value reduces as the oil becomes contaminated. The depth of modulation produced by

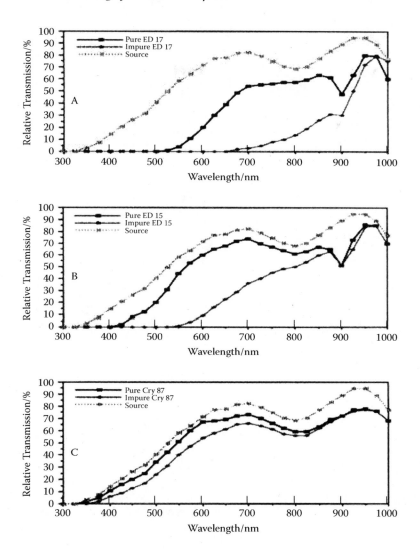

FIGURE 5.2.1.1
Transmission spectra for a white light source through air and absorption due to pure and impure oils: (a) Edwards 17; (b) Edwards 15; (c) Crylin 87. (From Khandaker, I., Glavas, E., and Jones, G.R., *Meas. Sci. Technol.*, 4, 1993. With permission.)

the contamination is given by

$$\frac{\nabla x}{x_0} = \frac{\text{Chromaticity of pure sample} - \text{Chromaticity of impure sample}}{\text{Chromaticity of pure sample}} \times 100\%$$

(5.2)

$(\nabla x/x_0)$ is different in each case.

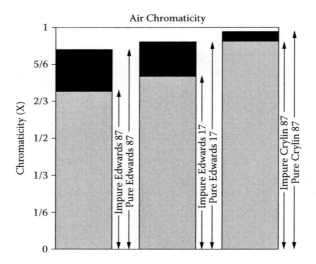

FIGURE 5.2.1.2

Values of chromatic parameter x for three different oil types in their pure and impure states. (From Khandaker, I., Glavas, E., and Jones, G.R., *Meas. Sci. Technol.*, 4, 1993. With permission.)

5.2.2   Aircraft Fuels Discrimination

5.2.2.1   *Single-Phase Fuel Types*

There are a number of different fuel grades that are used by aircraft. Although these fuels contain similar hydrocarbon chains, they have slightly different spectral characteristics (Barroqueiro and Jones, 2004). The spectral signatures of three fuels used in the aviation sector (AVGAS, AVCAT, and AVTUR) as well as air are shown in Figure 5.2.2.1. There are differences in the signatures of these fuels and air within the wavelength range 540–710 nm and also between 900–1000 nm. These are of a distributed nature and not confined to a few spectral lines. To capitalize on the identifiable differences within the two spectral ranges, an optical source consisting of a white-light LED with strong spectral emissions in the blue part of the spectrum, augmented with an infrared LED with a peak optical output at 950 nm, was used in the monitoring system. The light was transmitted via multimode optical fibers to a reflective probe, which was immersed into the various fuels, and the modulated signal was delivered via another set of optical fibers to a tristimulus detection unit.

The outputs from the three detectors of the tristimulus system are processed to produce values of H, L, S (see Chapter 1) for the various fuels. Despite the apparent similarities in the spectral characteristics over a large proportion of the spectra, there are unique and distinct differences in the H and S values (Figure 5.2.2.1b). These differences are significant, including the difference between air and AVTUR despite the similarity of their spectra.

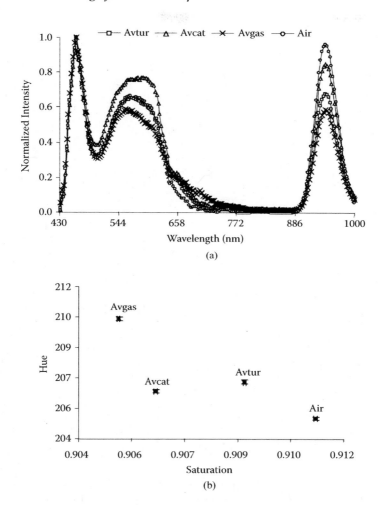

FIGURE 5.2.2.1

Chromaticity of aviation fuels: (a) transmission spectra for air, AVTUR, AVCAT, AVGAS, fuels; (b) *H* and *S* values for the transmission spectra in air, AVTUR, AVCAT, AVGAS fuels; (c) *H* and *S* values for different contents of water in AVGAS fuel. (From Barroqueiro, S. and Jones, G.R., *Meas. Sci. Technol.*, 15, 5, 2004. With permission.)

The high saturation values (> 0.9) are due to the biased nature of the white LED light source.

5.2.2.2   Water/Fuel—Two-Phase Fluids

Agitated mixtures of water and fuel (AVGAS) constitute a two-phase, emulsified fluid. The *H*, *S* coordinates for mixtures with various proportions of water in the fuel are different (Figure 5.2.2.1c, error bars indicating

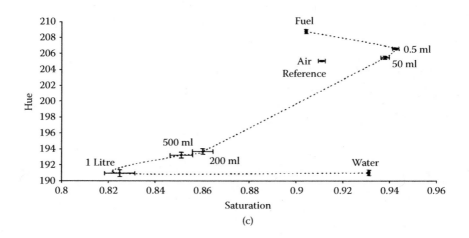

FIGURE 5.2.2.1
(Continued).

inhomogeneity of various test samples). Chromatic *H* values and their standard deviation vary inversely as the water content in the range 0.5 ml–1 l.

5.2.3   Thickness of Immiscible Layers

An opposite case to that of the thoroughly mixed liquids considered in Section 5.2.2 is the case of a thin layer of one liquid floating on the surface of another liquid. An example is a 100-µm layer of kerosene floating upon water (Smith et al., 1994). Monitoring the thickness of the thin layer of kerosene can be facilitated by mechanically amplifying the layer thickness by drawing the liquid into an inverted funnel (Figure 5.2.3.1a). As a result of the gradually reducing cross-sectional area of the funnel, the liquid layer thickness increases as the liquid is drawn into the funnel, making it more readily detectable. A chromatic system for monitoring the thin layer uses a tungsten halogen source, distimulus chromatic detection, and optical fiber transmission of the light to and from the sensor.

Some typical test results obtained with such a system are shown in Figure 5.2.3.1b for a 5-mm-thick kerosene layer on water and in Figure 5.2.3.1c for a 5.5-mm thick mineral oil layer on water. The results are in the form of dominant wavelength versus vertical displacement of the sensing unit (Figure 5.2.3.1a) in the liquid. In each case, as the sensing unit is withdrawn the optical transmission medium between transmit and receive optical fibers changes from water through kerosene or mineral oil to air. Consequently, the dominant wavelength value changes in accordance with the transmission medium. For example, with kerosene, the dominant wavelength changes from ~741 nm (water) via 747 nm (kerosene) to 749 nm (air). The thickness of the kerosene layer is determined from the positions of the abrupt changes in dominant wavelength at 0.65 mm on the vertical displacement scale

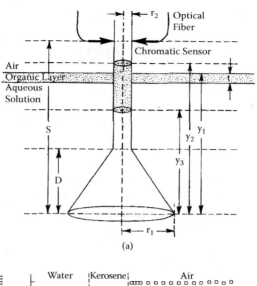

(a)

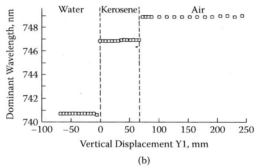

(b)

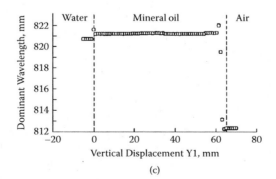

(c)

FIGURE 5.2.3.1
Chromatic optical fiber monitoring of thin immiscible layers: (a) schematic of inverted funnel transducer; (b) dominant wavelength changes for air, kerosene, and water; (c) dominant wavelength for air, mineral oil, and water. (From Smith, R., Spencer, J.W., Jones, G.R., Lightfoot, J., and Dean, E., *IEE Proc. Optoelectron.*, 141, 4, 1994. With permission.)

corrected for the geometric amplification (Smith et al., 1994). A similar procedure applied to Figure 5.2.3.1c allows the mineral oil layer thickness to be determined (the absolute values of dominant wavelength in the kerosene and mineral oil tests are different due to changes made to the system).

The accuracy of this type of chromatic sensing system is about 140 μm with a dynamic range of 0.3–20-mm layer thickness (Smith et al., 1994).

5.2.4   Chemical Indicators

The use of chemical indicators for detecting specific groups of chemicals or conditions (e.g., pH indicators and cobalt chloride moisture detectors) by changes in color is well established. Utilizing such indicators and chemical specific dyes in combination with optical- fiber sensing and chromatic processing provides a means for monitoring liquids on an industrial scale and continuously on line. Three examples of such optical fiber-based chromatic sensing are described: solvatochromatic dyes, pH indicators, and cobalt chloride solution.

5.2.4.1   *Water in Organic Solvents*

Solvatochromatic dyes (Russell et al., 1996) change color in response to a liquid in which they are dissolved. The change is dependent upon the polarity of the solvent and is used to indicate its composition. One example is a solvatochromatic dye for the detection of water in an organic solvent (butyl acetate). The dye is immobilized through covalent bonding on to a silica based optical fiber. The presence of water in butyl acetate causes the dye to change color, which is detectable with chromatic monitoring. Figure 5.2.4.1a shows the change in chromaticity produced in such a sensor for different levels of water in butyl acetate, on an x:y chromatic diagram (see Chapter 3, Section 3.2) The chromatic detection utilized a tungsten halogen source and tristimulus detection.

The probe is able to respond to small changes in water content in a reversible and continuous manner (Figure 5.2.4.1b).

5.2.4.2   *Optical Fiber Chromatic pH Sensing*

Optical fiber-based chromatic sensing of liquid acidity has been demonstrated using a paper base impregnated with a pH chemical indicator. As this sensing element needs to be exposed to the liquid under investigation, care should be taken to reduce the possibility of fouling the sensing element and the optical cavity. The pH indicator element is therefore housed in an enclosure with small access ports to prevent the ingress of any particulates, which could contaminate the sensing material and cavity (Figure 5.2.4.2a) (Rallis et al. 2005).

A dominant wavelength-pH calibration curve for a fiber optic pH monitoring unit is shown in Figure 5.2.4.2b (Rallis et al. 2005). This was acquired

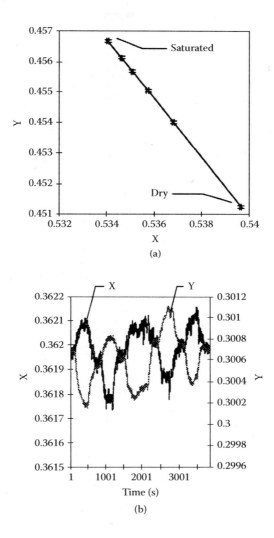

FIGURE 5.2.4.1
Chromatic optical fiber sensing with solvatochromatic dye: (a) chromatic *x, y* changes in butyl acetate when exposed to water free and saturated solvent; (b) changes in chromatic values *x* and *y* to repeated exposure of solvent with different water content. (From Russell, P.C., Jones, G.R., Crowther, D., Jones, M., Ahmed, S.U., and Huggett, P., in *Proc. Opto '96*, ACS Orgs. GmbH, Wunstorf, Germany, 1996, 145–150. With permission.)

using a distimulus monitoring system (see Chapter 3, Section 3.2) via a 30 m length of optical fiber. The indicator element can gradually bleach with resident time in a liquid but the sensor performance remains unaffected by continuous use over a period of several months (see Chapter 9, Section 9.3.2.1).

pH Sensitive
Element

Fiber Optic
Cable

Ports for
liquid
particles
filtering

O-Ring Seal

(a)

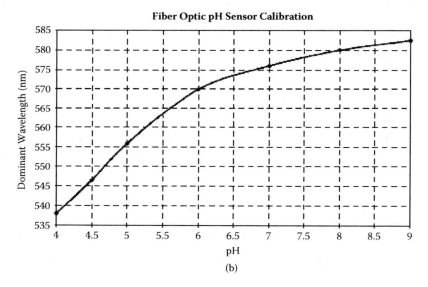

(b)

FIGURE 5.2.4.2
Optical fiber-based chromatic pH probe unit: (a) structure of probe; (b) dominant wavelength versus pH calibration curve. (From Rallis, I., Deakin, A., Spencer, J.W., and Jones, G.R., *17th Int. Conf. on Optical Fibre Sensors, Proc. of SPIE*, 5855, 2005. With permission.)

5.2.4.3   Cobalt Chloride Temperature Sensing

The spectrum of polychromatic light transmitted through an aqueous solution of cobalt chloride varies with temperature (e.g., Rallis, 2004). Such solutions contained in a small phial attached to the tip of an optical fiber can be used for temperature sensing. The chromatic change produced by such spectral variations may be calibrated to produce a temperature sensor for online monitoring. The variation of dominant wavelength (determined with a distimulus system) with temperature (Figure 5.2.4.3) is a sigmoid curve (see Chapter 3, Section 3.2, Figure 3.3). The most sensitive response region of the sensor can be tuned by varying the gains of the chromatic processors

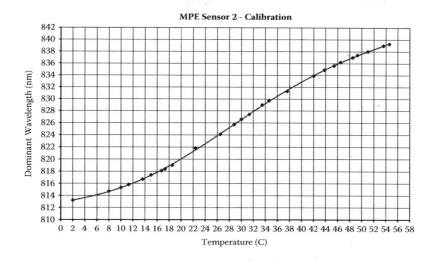

FIGURE 5.2.4.3
Dominant wavelength versus temperature for cobalt chloride. (From Rallis, I., Deakin, A., Spencer, J.W., and Jones, G.R., *17th Int. Conf. on Optical Fibre Sensors, Proc. of SPIE*, 5855, 2005. With permission.)

with respect to each other and without physically needing to change the sensor head itself.

5.3 Remote CCD Chromatic Sensing of Liquid Level

Miniature CCD cameras may be deployed in conjunction with space-domain chromaticity for monitoring liquid levels via a number of manifestations (Barroqueiro and Jones, 2004). These methods involve illuminating the liquid surface with three remotely situated LEDs, which are geometrically separated from each other and in a horizontal plane above the liquid surface. Reflections of light from each of the three LEDs off the liquid surface are monitored via images captured with the CCD camera. The geometric position of the light associated with each LED varies with the separation of the liquid surface from the CCD camera.

One manifestation of such a system is shown in Figures 5.3.1a and 5.3.1b. The geometric separation of the images of each of the three LEDs increases as the liquid level rises closer to the CCD camera (Figure 5.3.1b). The liquid level may be determined by deploying space-domain chromatic processing of the reflected images of the LED light, which can be achieved in several ways. Two examples are the deploying of space-domain chromatic filters (*R, G, B*) to monitor

  (a) radial displacement of an LED image (Figure 5.3.2a);
  (b) radial displacement of the illuminated meniscus (Figure 5.3.2b).

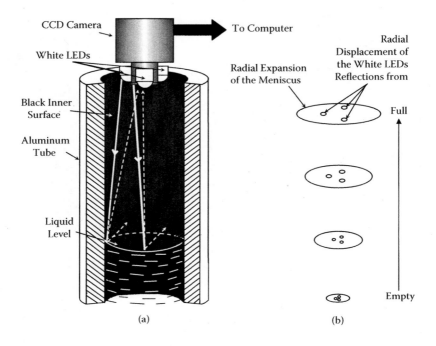

(a)   (b)

FIGURE 5.3.1
CCD camera-based liquid level monitor: (a) structure of the monitor unit; (b) schematic of image changes. (From Barroqueiro, S.A.B., 2002, *Chromatic Sensors for Aircraft Fuel Systems*, Ph.D. thesis, University of Liverpool.)

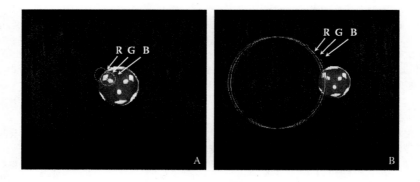

FIGURE 5.3.2
Processed CCD images showing *R*, *G*, *B* chromatic space filters: (a) *R*, *G*, *B* addressed LED reflection; (b) *R*, *G*, *B* addressed meniscus. (From Barroqueiro, S.A.B., 2002, *Chromatic Sensors for Aircraft Fuel Systems*, Ph.D. thesis, University of Liverpool.)

(The clarity of the images shown in Figures 5.3.2 was obtained after preprocessing the captured image (Barroqueiro, 2002).

In each case the R, G, B filter outputs are transformed into chromatic H, L, S coordinates, where H represents the dominant position of the light, L its strength, and S its spread.

The liquid level is determined from a calibration curve relating chromatic H to the level, an example of which is given in Figure 5.3.3a for the meniscus deployment of the chromatic filters (Figure 5.3.2b). Hysteresis effects associated with rising and falling liquid levels are of the order of 0.09% (Barroqueiro, 2002). A similar H-liquid level calibration curve is obtained with the deployment of the R, G, B filters on the LED image (Figure 5.3.2a) but with a hysteresis effect of approximately 2.7%.

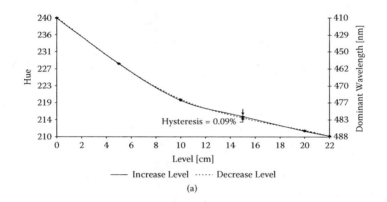

(a)

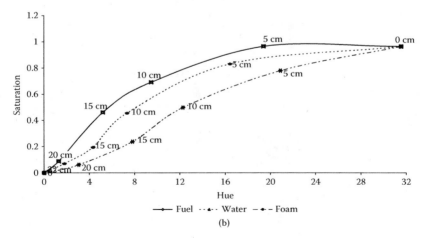

(b)

FIGURE 5.3.3
Chromatic diagrams for CCD-derived data: (a) H and dominant wavelength versus liquid level for the LED reflections; (b) H versus S diagram for water, fuel, and foam for the meniscus case. (From Barroqueiro, S.A.B., 2002, *Chromatic Sensors for Aircraft Fuel Systems*, Ph.D. thesis, University of Liverpool.)

Additional information is available from the other space-domain chromatic parameters S, L. A distinction can be made between different liquid conditions, e.g., in discriminating between fuel, water, and foam. Each of these three conditions has a unique H-S signature leading to three different characteristic curves on an H-normalized S diagram (Figure 5.3.3b). The level and nature of the liquid interface are therefore distinguishable from this diagram.

The CCD–space chromaticity approach has been extended not only to determine the depth and type of liquid but also to distinguish immiscible liquids. The probe geometry is modified so that it incorporates a Perspex sheet to divide the monitoring cavity into two compartments—one open to receive the liquid, the other sealed from the liquid and containing only air. Two white-light LEDs are used to transmit light through the Perspex elements. A third LED is placed on the cavity open to the liquid and the light is projected onto the surface of the liquid (Figure 5.3.4). A CCD camera is used to capture images of the cavity interior and LED reflections shown in Figures 5.3.5a and 5.3.5b corresponding to different thicknesses of immiscible layers of fuel on water. The deployment of three space-domain chromatic

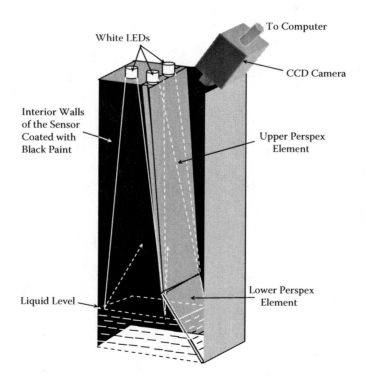

FIGURE 5.3.4
CCD-based chromatic level and immiscible layer sensing unit. (From Barroqueiro, S.A.B., 2002, *Chromatic Sensors for Aircraft Fuel Systems*, Ph.D. thesis, University of Liverpool.)

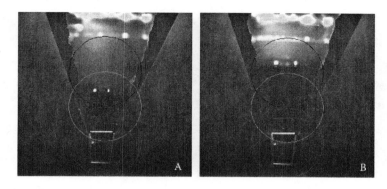

FIGURE 5.3.5
CCD images and space chromatic filters: (a) 1 cm of fuel on water; (b) 0.5 cm of fuel on water. (From Barroqueiro, S.A.B., 2002, *Chromatic Sensors for Aircraft Fuel Systems*, Ph.D. thesis, University of Liverpool.)

filters is as shown. Preprocessing of the images was not required with such a deployment.

The liquid level and the thickness of the immiscible layer can be determined from an *H-S* chromatic diagram in which the data form characteristics similar to those in Figure 5.3.3b. Changes in the liquid level are predominantly indicated by changes in the *H* parameter (25 cm displacement producing a 216° change) (Barroqueiro, 2002). Increasing the layer thickness produces a decrease in the *S* parameter (~0.07 at a depth of 20 cm compared to a resolution of 0.01).

5.4 Discrete Chromatic Processing of Multisensors for Aircraft Inclination Indications

Aircraft fuel systems employ a multiplicity of sensors to monitor various parameters (pressure, flow, level, temperature, etc.) within fuel tanks and interconnecting pipes and pumps. A schematic of a model aircraft fuel tank is shown in Figure 5.4.1 along with the array of sensors deployed. A total of 18 sensors are deployed, all providing relevant detailed data about the fuel system condition.

Discrete chromatic processing (see Chapter 2, Section 2.5) can be applied to the outputs of the 18 sensors to provide additional information for dynamically tracking the attitude (roll and pitch angle) of an aircraft.

The deployment of the discrete chromatic processing for fusing the outputs of the system sensors (see Chapter 2, Section 2.5) involves overlaying triangular processors (*R, G, B*) on the sequence of the outputs from the 18 sensors (see Chapter 2, Figure 2.5.2,).

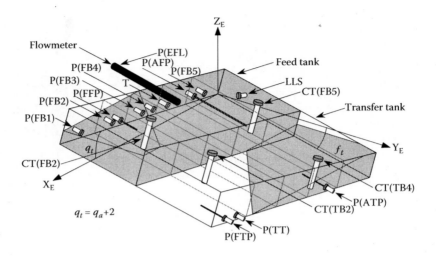

FIGURE 5.4.1
Schematic of an aircraft fuel tank system and sensor network. (From Anupriya, 2001, *Multi-Sensor Data Fusion for Aircraft Fuel Systems Using Chromatic Processing*, Ph.D. thesis, University of Liverpool.)

To highlight changes in aircraft attitude, only three sensor elements (S_{n1}, S_{n10}, S_{n18}) are retained, because the remaining sensors have been shown to be insensitive. For maximum sensitivity, these three sensor elements have been located at element positions 1 (S_{n1}), 10 (S_{n10}), and 18 (S_{n18}) (Figure 5.4.2a).

Each sensor output, V_i, is normalized with respect to the useful range of the sensor, i.e.,

$$V_i = \frac{V_i - V_{min_i}}{V_{max_i} - V_{min_i}} \tag{5.3}$$

where V_{min} and V_{max} are the minimum and maximum level signals, respectively. The outputs from the three processors (R, G, B) may then be chromatically transformed using the CIE *Lab* algorithms (see Chapter 3, Section 3.4, Equations 3.26 through 3.28). For this particular application, the reference values R_n, G_n, B_n are the tristimulus values of a reference signal corresponding to those for level flight with a pitch angle of $-3°$ and a roll angle of $0°$ (Figure 5.4.3; Anupriya, 2001).

The chromatic processing has been evaluated via tests simulating a standard aircraft flight profile consisting of six principal flight phases: engine start and taxiing, takeoff, climb, cruise, descent, and landing and taxiing.

The results of the chromatic mapping for the simulated flight are shown in Figure 5.4.3 as a 2-D *Lab* plot. A substantial number of experimental points are shown corresponding to each time instance of the model flight profile summarized on the insert box in Figure 5.4.3. Each phase of the model flight is identified on Figure 5.4.3 by a different symbol (X, O, etc.), and various

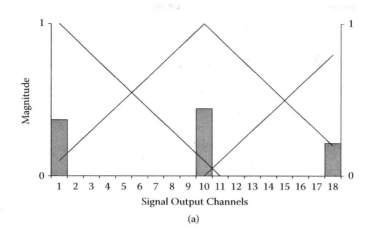

(a)

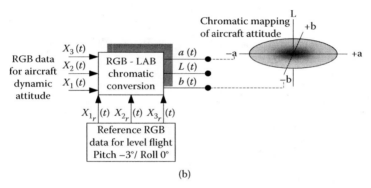

(b)

FIGURE 5.4.2
Discrete *Lab* chromatic processing: (a) Superposition of *R, G, B* detectors on three output signals; (b) normalized *R, G, B* processing. (From Anupriya, 2001, *Multi-Sensor Data Fusion for Aircraft Fuel Systems Using Chromatic Processing*, Ph.D. thesis, University of Liverpool.)

aircraft attitudes (roll, pitch) are further discriminated by variants of these symbols (e.g., X – Roll 0°; X – Roll +/– 30°).

Normal flight (pitch angle –3°, roll angle 0°) conditions, regardless of fuel consumption, are represented by the origin ($a = b = 0$) throughout the flight. Changes in roll angle (–30° to +30°) at a pitch of –3° are apparent as a displacement of the operating point linearly from the origin with a gradient of approximately –45° to the *a, b* axis (AB in Figure 5.4.3). Changes in pitch angle (–16° to +14°) at a roll angle of 0° appear as displacements from the origin, approximately orthogonal to the roll angle variation locus (CD in Figure 5.4.3). For a fixed pitch angle of –16°, a superimposed roll leads to the roll angle characteristic line being displaced laterally from the –3° pitch angle characteristic (EF in Figure 5.4.3). Various combinations of roll and pitch angles therefore lead to the instantaneous aircraft attitude being represented unambiguously by a point on the chromatic *a–b* map.

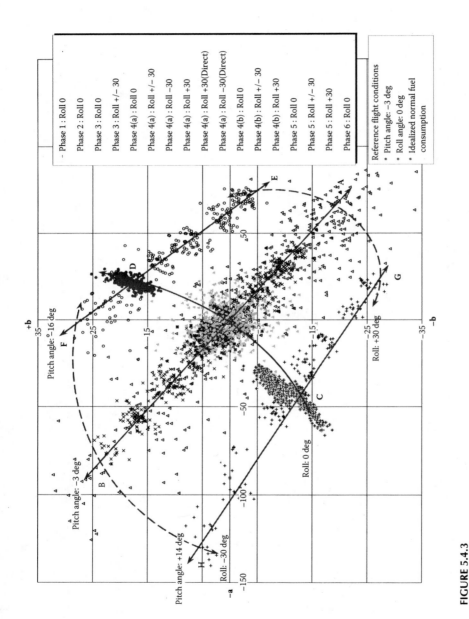

FIGURE 5.4.3

(See color insert following page 18). CIE *Lab* chromatic diagram of aircraft attitude variation during various phases of flight simulation. (From Anupriya, 2001, *Multi-Sensor Data Fusion for Aircraft Fuel Systems Using Chromatic Processing*, Ph.D. thesis, University of Liverpool.)

The results shown in Figure 5.4.3 are subject to the effect of noise on the transducer outputs leading to a degree of scattering on the operational points. The noise was more noticeable for roll rotations in the range 0° to −30° due to the lower sensitivity of the transducers to negative roll rotations. Furthermore, as the fuel is consumed from the tanks, the noise level also increases because of more vigorous fuel mixing produced by fuel movement.

5.5 Summary

Chromatic monitoring of industrial liquids can be deployed with either optical fiber or remote sensing as well as for fusing the outputs from a multiplicity of conventional, electronic sensors.

Optical fiber sensing based upon wavelength-domain chromaticity provides a means for

- monitoring the degradation of liquids such as oils in vacuum pumps;
- distinguishing between different types of aviation fuels;
- detecting water in fuels;
- determining the thickness of immiscible layers of liquids;
- addressing various color-based sensing materials, such as solvatochromatic dyes, pH indicators, etc., for detecting impurities or the state of a liquid.

CCD camera monitoring based upon space-domain chromaticity can be used for

- monitoring liquid levels;
- distinguishing between fuel, water, and foam;
- determining the thickness of immiscible layers.

Discrete chromatic processing of the outputs of a multiplicity of fuel system sensors enables the attitude of an aircraft in flight to be tracked.

References

Anupriya (2001). *Multi-Sensor Data Fusion for Aircraft Fuel Systems Using Chromatic Processing*, Ph.D. thesis, University of Liverpool.

Barroqueiro, S.A.B. (2002). *Chromatic Sensors for Aircraft Fuel Systems*, Ph.D. thesis, University of Liverpool.

Barroqueiro, S.A.B. and Jones, G.R. (2004). Chromatic single-point sensor for aircraft fuel systems, *Meas. Sci. Technol.*, 15(5), 814–820.

Jones, G.R., Jones, R.E., and Jones, R. (2000). Multimode optical fiber sensors, in *Optical Fiber Sensor Technology*, Grattan, L.S. and Meggitt, B.T., Eds., Kluwer Academic, Dordrecht, The Netherlands, pp. 1–78.

Khandaker, I., Glavas, E., and Jones, G.R. (1993). A fibre-optic oil condition monitor based on chromatic modulation, *Meas. Sci. Technol.*, 4, 608–613.

Rallis, I. (2004). *Intelligent Chromatic Fibre Optic Sensors and Monitoring Systems for Enhancing Useful By-Products from Anaerobic Digestion*, Ph.D. thesis, University of Liverpool.

Rallis, I., Deakin, A., Spencer, J.W., and Jones, G.R. (2005). Novel sensing techniques for industrial scale bio-digesters, *17th Int. Conf. on Optical Fibre Sensors, Proc. of SPIE*, 5855, 110–113.

Russell, P.C., Jones, G.R., Crowther, D., Jones, M., Ahmed, S.U., and Huggett, P. (1996). A chromatic sensor for detecting water in organic solvents, *Proc. Opto '96*, 145–150. ACS Orgs. GmbH, Wunstorf, Germany.

Smith, R., Spencer, J.W., Jones, G.R., Lightfoot, J., and Dean, E. (1994). Fibre optic probe for determining the thickness of immiscible layers, *IEE Proc. Optoelectron.*, 141(4).

6

Chromatic Monitoring for Broadband Interferometry and Polarimetry

C.A. Egan, G.R. Jones, and C.D. Russell

CONTENTS

6.1   Introduction

Two monochromatic light waves can interact with each other to produce
interference patterns from which small distances of the order of the optical
wavelength can be determined (interferometry). A light wave propagating
through some materials can have the geometric plane in which it oscillates
modified by the structure of the material (polarimetry). This in turn can
be affected by external factors such as temperature, mechanical stress, and
magnetic field. Monitoring the polarized state of the light can therefore be
used for tracking changes in such parameters.

Such phenomena have conventionally been deployed with highly monochromatic beams of light, but they are also wavelength dependent. This leads to the
possibility of using polychromatic beams for monitoring via such processes
and which can be visually observed. Figure 6.1.1 shows the visualization of
some classical patterns of polychromatic fringes on a diaphragm, the form of
which yields an indication of the stress distribution across the diaphragm.

An advantage of a polychromatic approach is that more information becomes available. The disadvantage hitherto has been the complexity of that
information. For example, with monochromatic light it is difficult to distinguish between different fringes formed on a diaphragm, such as that shown
in Figure 6.1.1. With polychromatic light, changes in the coloration of groups
of fringes (fringe orders) enable such discrimination to be made. Quantifying such complicated polychromatic effects caused by changes in the wave
properties (e.g., polarization, interference) in a precise, robust, and cost-
effective manner has been regarded as difficult, but is achievable with chromatic techniques.

Interferometric and polimetric principles are briefly discussed with regard
to broadband (polychromatic) deployment. Examples of chromatic monitoring

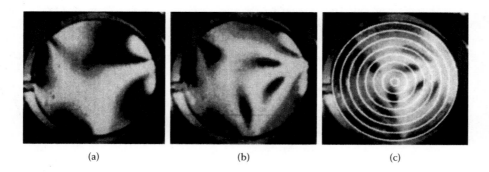

| (a) | (b) | (c) |

FIGURE 6.1.1
(See color insert following page 18). Images of a photoelastic diaphragm with stresses
produced by various pressure differences across the diaphragm: (a) 20 lb/in.^{2}, (b) 40 lb/in.^{2},
(c) eight annular rings used for 2-D analysis (Ahmed, S. U., Intelligent Remote Chromatic Processing, Ph.D. thesis, University of Liverpool, U.K., 1998).

using these phenomena are given including the monitoring of thin films, composition of optically active chemicals (e.g., sucrose), mechanical stress (via photoelastic materials), temperature (via thermochromic materials), and electric current (via magneto-optic materials).

6.2 Broadband Optical Interferometry

6.2.1 Basic Principles

Interferometry involves the interaction of a number of light beams with each other to produce interference patterns consisting of a series of bright and dark fringes (Grattan and Meggitt, 1995; Guenther, 1990). Arrangements for producing such interference are known as interferometers. They may be classified according to whether two or more beams are involved.

An example of a two-beam interferometer (Michelson) is shown in Figure 6.2.1.1a. A signal beam (2) from a movable mirror M_s combines with a reference beam (1) from a fixed mirror M_r via a beam splitter to provide interference between the two beams at an optical detector D.

An example of multiple interference (Fabry-Perot cavity) is shown in Figure 6.2.1.1b. Beams E_{r0}, E_{r1}, E_{r2}, etc., produced by partial reflection or transmission through boundary M_1 and multiple reflection from the second boundary M_2 of the cavity interact to form an interference pattern.

Interferometers which utilize broadband spectral sources such as LEDs or halogen lamps are low coherence interferometers (Meggitt, 1995). The sensor intensity profile $I(k)$ is a function of the wave number, $k = 2\pi/\lambda$ (e.g., Figure 6.2.1.2a).

$$I(\lambda) = I_0 \exp\left[-\left(\frac{k - k_0}{\sigma/2}\right)\right] \tag{6.1}$$

where k_0 is the central wave number, $\sigma/2$ is the half-width, I_0 is the optical power at k_0.

There are two basic schemes for processing the signals from broadband interferometers which are spectral- and phase-domain processing.

In the spectral domain, the output power is given by (Meggitt, 1995) the product of the spectral width (Equation 6.1) and a periodic function representing the fringe

$$I(k) = I_0 \exp\left[-4\left(\frac{k - k_0}{\sigma}\right)^2\right][1 + \cos(\varphi(k, \delta))] \tag{6.2}$$

where δ is the optical path difference in the interferometer cavity, φ is the phase angle, e.g.,

$$\varphi = (4\pi n L)/\lambda_0 \tag{6.3}$$

where n is refractive index, L is geometric path.

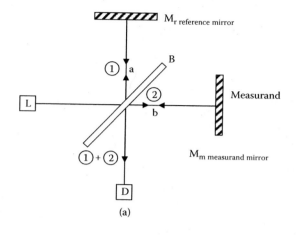

(a)

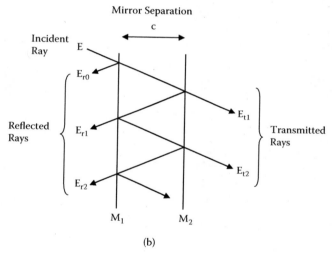

(b)

FIGURE 6.2.1.1

Some basic interferometer arrangements: (a) two-beam Michelson interferometer. (From Jones, G.R., Jones, R.E., and Jones, R., in *Optical Fiber Sensor Technology*, Kluwer Academic, Dordrecht, The Netherlands, 2000, 1–77. With permission; (b) illustration of low-finesse Fabry–Perot cavity formed between two parallel and partially reflecting mirrors, M_1 and M_2, showing transmitted and reflected ray sets (From Meggitt, B.T., in *Optical Fiber Sensor Technology*, Chapman & Hall, London, 1995, 269–312. With permission.)

Equation 6.2 represents an output fringe pattern with respect to wave numbers as shown in Figure 6.2.1.2(b)

In the phase domain, the reference signal may be time-varied. For instance, with the Michelson interferometer (Figure 6.2.1.1a) the reference mirror M_r is vibrated so that the reference path length oscillates. The output power is

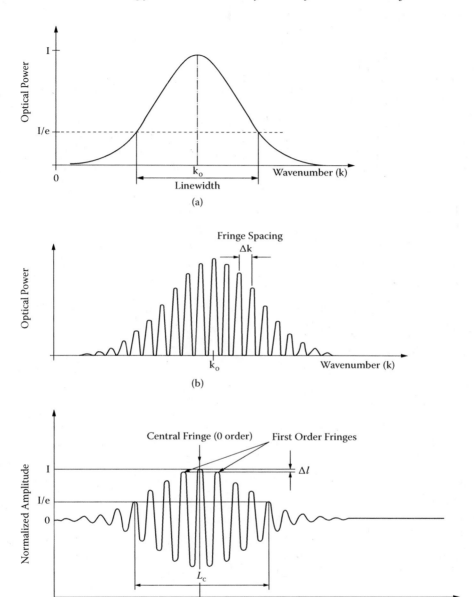

FIGURE 6.2.1.2
Broadband interferometer signal: (a) Gaussian spectral intensity profile from a typical low coherence source (e.g., LED); (b) idealized cosine spectral fringes produced when using a spectrometer as the processing interferometer (spectral domain). (From Meggitt, B.T., in *Optical Fiber Sensor Technology*, Chapman & Hall, London, 1995, 269–312. With permission.); (c) phase domain broadband interferometer signal. (From Meggitt, B.T., in *Optical Fiber Sensor Technology*, Chapman & Hall, London, 1995, 269–312. With permission.)

then given by (Meggitt, 1995)

$$I(k,\delta_1,\delta_2) = I_0\left[1 + \exp[-((\delta_1 - \delta_2)\sigma/2)^2]\cos(k(\delta_1 - \delta_2))\right] \qquad (6.4)$$

where $\delta_1 - \delta_2$ is the optical path difference (nL) between a signal and a reference beams. The output signal is of the form shown in Figure 6.2.1.2c.

The significance of Equations 6.3 and 6.4 is that they relate the light intensity variation (I/I_0) to the optical path difference δ, and hence the geometric length L and the refractive index n. As L or n varies, (I/I_0) passes through a series of maximum and minimum values (fringes). Hence, a count of the number of fringes observed yields the change in optical path in terms of the wavelength.

6.2.2 Chromatic Processing of Wavelength-Domain Interference Pattern

In broadband interferometry, the connection between the spectral and phase domains (Section 6.2.1) may be shown by plotting the optical fringe intensity as a function of both the optical path length, nL, and the optical wavelength on a 3-D diagram (Figure 6.2.2.1; Jones et al., 1994). This shows fringes produced by a Fabry–Perot cavity (Figure 6.2.1.1b, Section 6.2) for light of three different wavelengths increasing in the order λ_1, λ_2, λ_3. For a gap width, nd, the figure shows that there are phase differences between the fringe patterns produced at each of the three wavelengths. Consequently, the interference with polychromatic light will have spectral signature variations shown by the shaded regions in Figure 6.2.2.1 at each of the three

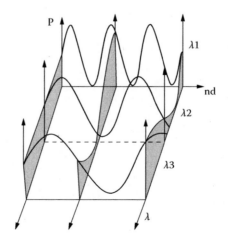

FIGURE 6.2.2.1
Polychromatic optical interference (interference fringes at three wavelengths (λ_1, λ_2, λ_3) as a function of the thin film thickness and shaded areas indicate the optical spectrum at each of three arbitrary film thickness). (From Jones, G.R., Russell, P., and Khandaker, I., *Meas. Sci. Technol.*, 5, 1994. With permission.)

nd locations. The implication is that the spectrum of the light observed changes with the width *nd* of the gap. This is the basis of the colors produced in nature by thin films such as the scales on the wings of some butterflies (e.g., Tilley, 2000).

The spectral signatures of the form shown in Figure 6.2.2.1 produced by polychromatic interference may be addressed with three chromatic processors that are nonorthogonal in the wavelength domain (see Chapter 1). Each spectrum that corresponds to a particular gap width, *nd*, is then defined in terms of particular values of the chromatic coordinates. The chromatic algorithms employed by Jones et al. (1994) correspond to the tristimulus ones of Equations 3.24 and 3.25 (see Chapter 3, Section 3.3).

$$x_1 = \frac{A[ai_1 - c(i_1 + i_2 + i_3)]}{(i_1 + i_2 + i_3)} \tag{6.5}$$

$$x_2 = \frac{B[bi_2 - c(i_1 + i_2 + i_3)]}{(i_1 + i_2 + i_3)} \tag{6.6}$$

where i_1, i_2, i_3 are the outputs of detectors *R*, *G*, *B* (R_0, G_0, B_0, equations 3.24, 3.25) and *A*, *B*, *a*, *b*, *c* (= *m*) are defined in Chapter 3, Section 3.3. The monitoring range and sensitivity are controlled via the weighting factors *A*, *a*, *B*, *b*, *c*.

6.2.2.1 Monitoring Semiconductor Thin Films

The principles described in Section 6.2.1 may be used for monitoring the etching of materials (e.g., SiN_3) in the production of semiconductor devices (e.g., Jones et al., 1994). Figure 6.2.2.2a shows a calibration curve relating the value of each of the chromatic parameters x_1, x_2 (Equations 6.5 and 6.6) to the film thickness. (In this case, the constants *A*, *B*, *a*, *b* were set to unity and *c* = 0 yielding $x_1 = x$, $x_2 = y$.)

The results show that both *x* and *y* vary cyclically but out of phase with each other over a thickness ranging from ~20 nm to 250 nm. This enables in principle ambiguities that might arise via the nonmonotonic nature of each chromatic parameter alone to be overcome. Consequently, the thickness of the film can be determined from a look-up table of *x*, *y*, *nd*.

The *x*, *y* variation may alternatively be displayed in chromaticity space to yield the convoluted curve shown in Figure 6.2.2.2b (time increasing as indicated by the arrow). Displaying the results in this chromatic form provides two advantages:

- It highlights a point of ambiguity (~0.48, 0.405) where the convolute crosses itself. For the time-carrying thickness example, the ambiguity may be resolved by determining the gradient of the curve at the point of intersection. Alternatively, the third chromatic parameter, L, may be applied to provide the discrimination.

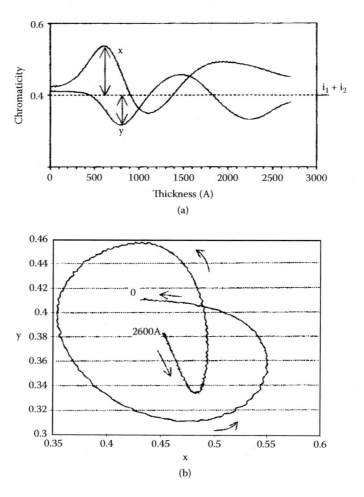

FIGURE 6.2.2.2
Chromatic signals for different thickness of a thin film ($x = i_1/(i_1 + i_2 + i_3)$; $y = i_2/(i_1 + i_2 + i_3)$; $A = B$ $= a = b = 1$, $c = 0$, Equations 6.3 and 6.4): (a) variation of the chromatic parameters x, y with film thickness, nd; (b) x:y chromatic map showing the convolute corresponding to the time varying increase of the thin film thickness (arrows indicate increasing time). (From Jones, G.R., Russell, P., and Khandaker, I., *Meas. Sci. Technol.*, 5, 1994. With permission.)

- It highlights undesirable effects via departures from the calibration convolute curve. For example, changes in refractive index (hence composition) of the thin film affect the curve (Equation 6.3) exasperated by the dependence of n, upon optical wavelength.

For this particular example, the film thickness measurement range was 250 nm with a resolution of 2 nm. The range is determined by the coherence length being greater than the path length differences, which is estimated as ~1 μm for this particular case (Jones et al., 1994).

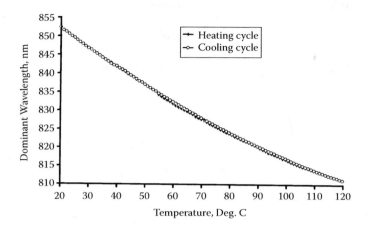

FIGURE 6.2.2.3
Dominant wavelength versus temperature for a chromatically addressed cavity temperature sensor. (From Jones, G.R., in *High Voltage Engineering and Testing*, IEE, London, 2001, 601–634. With permission.)

6.2.2.2 Chromatically Addressed Fabry–Perot Pressure and Temperature Sensors

By suitable design a Fabry–Perot Cavity may be made either pressure- or temperature-sensitive. By addressing the cavity with broadband light and monitoring the output as described in Section 6.2, chromatic pressure and temperature transducers can be produced.

Such chromatic pressure and temperature transducers have been used for monitoring conditions within high voltage, gas-blast circuit breakers (Jones, 2001; Messent et al., 1997). The variation of H (dominant wavelength) with temperature of such a transducer is shown in Figure 6.2.2.3, a temperature difference of 100°C producing a dominant wavelength change of $H = 40°$ with little hysteresis effects.

6.2.3 Chromatic Processing of Phase-Domain Interference Patterns

6.2.3.1 Phase-Domain Interference Signals

The optical interference signals produced by several interfaces separated by short distances may be addressed in the phase domain with a Michelson interferometer (Figure 6.2.1.1a) where m_s represents the interfaces, and m_r the moving reference, oscillated to scan the relevant optical path lengths. The interferograms are only produced when the optical path length to a particular interface matches that of the reference. As the reference moves, interferograms are produced sequentially with each interface in the time domain through the depth of the sample. The final interference pattern may be regarded as the addition of a series of interferograms derived from reflections from each interface. This addition of interferograms produces a simple pattern if the subsurface interfaces are spaced sufficiently far apart. However, when the separation of the interfaces reduces below the coherence length of the light source (Section 6.2.1), the component interferograms overlap to produce more complex patterns.

Computer-simulated diagrams of the output signal from a detector D of the interference, as a function of path length, can be produced to show the complexity of the patterns obtained (Russell et al., 2005). Figure 6.2.3.1 shows such simulations for a two-interface structure, each interface with a dissimilar reflectance. The simulations considered separations where the carrier cosine wave of the interferograms from each interface were either in phase, or one quarter of the wavelength out of phase. These two conditions would lead to either constructive or destructive interference (Section 6.2.1) between the two component interferograms. Figure 6.2.3.1a is the result for two interfaces separated by a distance of 26 µm and Figure 6.2.3.1b for a separation of 26.325 µm (the wavelength and coherence length used in the simulations were 1.3 µm and 28 µm, respectively). Each figure shows the component interferograms from each interface (ii) as well as the overall interference pattern seen at the detector D (i). The amplitude of each of the two component interferograms is different because of the difference in reflectance of the two interfaces. The form of the subsequent interferogram is markedly different as a result of the separation leading to either constructive (a) or destructive (b) interference as shown by the combined interference patterns in Figure 6.2.3.1a,b. These interference effects can cause significant problems in determining interfaces separated by small distances and in interpreting the measurements, not least in the case where constructive interference causes the peaks of the component interferograms to move closer together and become difficult to resolve.

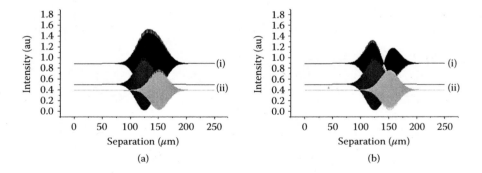

FIGURE 6.2.3.1
Interference between two interferograms (Interfaces with dissimilar reflectances: (i) resultant interferogram; (ii) component interferograms): (a) constructive interference; (b) destructive interference. (From Russell, C.D., Bullough, T.J., Jones, G.R., Krasner, N., and Bamford, K.J., in *Alt'03 International Conference on Advanced Laser Technologies: Biomedical Optics*, Vol. 5486, SPIE–International Society of Optical Engineering, Bellingham, Washington, 2003, 123–128. With permission.)

6.2.3.2 *Chromatic Analysis of Phase-Domain Interference Signals*

The decoding of interference signals of the form shown in Figure 6.2.3.1a,b to determine the separation of the two interfaces can be facilitated chromatically.

In this case, the Gaussian processors are shifted incrementally by their own width along the signal to produce a sequence of H, L, S values that are plotted as a function of position.

Three space-domain chromatic processors (R, G, B) are shown in Figure 6.2.3.2a along with a typical computer-simulated interference signal from a two-interface structure with narrow separation between the interfaces. This example highlights the difficulty in signal interpretation, as the peaks of the two component interferograms are not easily resolvable. Each chromatic processor had a half-width of 0.8 µm with peaks separated by 1 µm but are shown amplified in space by an order of magnitude in Figure 6.2.3.2a. By stepping the processors in space through the interference signal, values for the chromatic parameters H, L, S are obtained for each space step. The space variation in H, L, S through the signal of Figure 6.2.3.2a(i) is shown in Figure 6.2.3.2b(i). Two step-changes occurring in the H curve correspond to the position of the two interfaces in the simulation.

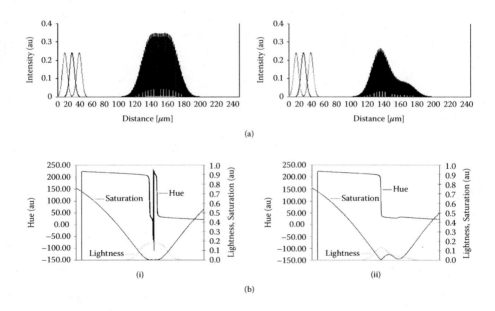

FIGURE 6.2.3.2
Examples of simulated interference signals and their chromatic signatures: (a) simulated signals and the R,G,B chromatic processors (the latter are shown x 10 FWHM); (b) Corresponding H,S,L values obtained by stepping R,G,B through the signals: (i) interfaces with equal reflectances; (ii) interfaces with dissimilar reflectances. (From Russell, C.D., Bullough, T.J., Jones, G.R., Krasner, N., and Bamford, K.J., in *Alt'03 International Conference on Advanced Laser Technologies: Biomedical Optics*, Vol. 5486, SPIE–International Society of Optical Engineering, Bellingham, 2003, 123–128. With permission.)

Figure 6.2.3.2a(ii) shows an interference signal from two interfaces with different reflectances so that the signal appears skewed. When addressed by time stepping with the three chromatic processors the space variations for H, L, S shown in Figure 6.2.3.2b(ii) is obtained. The location of the more strongly reflecting interface is indicated clearly by a substantial step in H (~150°) and the weaker reflecting interface by a smaller H step (~10°). In this case, a second minimum in the S curve (which can correspond to either minima or maxima in the signal) is clearly visible, corresponding to the position of the smaller H step confirming the location of the interface.

While this approach has been shown, both computationally and experimentally, to provide enhanced resolution of closely separated adjacent interfaces, it cannot explicitly confirm the presence of the aforementioned interference effects.

6.3   Broadband Optical Polarization

6.3.1   Basic Principles

A light wave is a transverse electromagnetic wave whereby its electric field vector is in a plane (x, y) perpendicular to the direction of propagation (z) (Figure 6.3.1.1.). The electric field vector may be defined in terms of its components (E_x, E_y) along the x, y directions by (e.g., Guenther, 1990)

$$\left(\frac{E_x}{E_{ox}}\right)^2 + \left(\frac{E_y}{E_{oy}}\right)^2 - \left(\frac{2E_x E_y}{E_{ox} E_{oy}}\right)\cos \nabla = \sin^2 \nabla \tag{6.7}$$

where E_{ox}, E_{oy} are maximum values, ∇ is the phase difference between E_{ox}, E_{oy}.

The direction of the electric vector E is the direction of polarization and the plane containing the vector is the plane of polarization (Figure 6.3.1.1).

Equation 6.7 allows three main types of polarization to be identified:

- The general case is that of elliptical polarization (OEe; Figure 6.3.1.1a) (Equation 6.7 is the equation of an ellipse)
- $\nabla = 0, N\pi$, leads to linear polarization (OA; Figure 6.3.1.1a) [Equation 6.7 becomes $(E_x/E_{ox}) = \pm (E_y/E_{oy})$]
- $E_{ox} = E_{oy} = E_o$, $\nabla = \pm\pi/2$, leads to circular polarization (OEc; Figure 6.3.1.1b) [Equation 6.7 becomes $(E_x)^2 + (E_y)^2 = E_o^2$]

In general, a collection of light waves will have individual waves with vectors orientated in different directions in the x, y plane and consequently are randomly polarized. Polarized light may be produced from these waves by their passage through a polarizing filter, which eliminates all but the waves with the required polarization. For example, the filter may only transmit

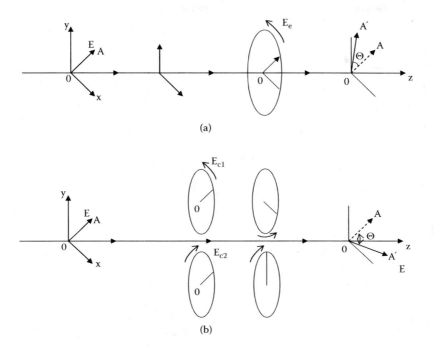

FIGURE 6.3.1.1
Propagation of polarized light along the z-axis: (a) birefringence; (b) optical activity.

waves whose electric vector is orientated along a particular inclination in the x, y plane (Figure 6.3.1.1) to provide linearly polarized light.

6.3.2 Polarization Effects in Materials

There are materials which can affect the state of polarization of light through which it propagates. The mechanism of interaction depends upon the nature and structure of the material (Guenther, 1990).

6.3.2.1 Birefringent Materials

Birefringence is a property of materials with structural anisotropy as a result of which the refractive index varies with the direction of the polarization of light (e.g., Guenther, 1990). Figure 6.3.1.1 shows light polarized along the direction oA propagating through a material along the z direction. The polarization of the wave may be resolved along two orthogonal directions, x, y, along which the refractive indices are a maximum and minimum, respectively, owing to the structural anisotropy of the material. As refractive index is an indicator of velocity, the two polarization components ox, oy travel with different velocities. This leads to polarized components developing a phase difference between each other at each point in the material. When the two linearly polarized components emerge from the material, they are recombined leading to an overall elliptical state of polarization. Thus, the linearly polarized input light has been converted into elliptical polarized light at the output. When viewed through a linear polarizing filter the x-y plane of polarization of the light appears to have been rotated through an angle (Figure 6.3.1.1a).

There are materials whose structural anisotropy may be varied by external influences such as mechanical stress (photoelasticity) and electric fields (electro optics) (Guenther, 1990).

6.3.2.2   Optically Active Materials

A linearly polarized beam of light may also be resolved into two circularly polarized components rotating contrary to each other (Guenther, 1990): E_{c1}, E_{c2} (Figure 6.3.1.1b). In materials that consist of a helical arrangement of molecules (Figure 6.3.2.1), the refractive index encountered by each of the contra-rotating wave components is different. Consequently, each wave propagates with a different velocity so that after a given distance of propagation there develops a phase difference between the two waves. On emerging from the material, the two circularly polarized waves recombine to form a linearly polarized wave, but one whose x-z plane if polarization is different from the incident wave (Figure 6. 3.1.1b). Materials which rotate the plane of polarization in this manner are said to be "optically active."

6.3.2.3   Magneto-Optic Materials

Magneto-optic materials produce similar but not identical light wave polarization effects to optically active materials, when subjected to a magnetic field. One major difference with magneto-optic materials is that the rotation

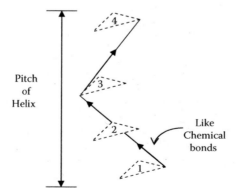

FIGURE 6.3.2.1
Molecular helix structure. (From Guenther, R.D., *Modern Optics*, John Wiley, New York, 1990, Figure 13.19. With permission.)

of the plane of polarization is independent of the direction of light propagation along the magnetic field direction. Thus, for a beam returning along its incident path, the effect is accumulative. With optically active materials the rotation depends upon the direction of light propagation. Consequently, there is no net change in polarization due to optical activity when a light beam is returned along its incident path in an optically active material (Ditchburn, 1958, p. 527; Guenther, 1990, p. 590).

In each of the foregoing three cases there is a phase difference produced between the two polarization components of the light beam (Guenther, 1990, p. 90; Ditchburn, 1958, p. 381), which leads to a rotation of the x-z plane of polarization through an angle (Guenther, 1990, pp. 550, 582)

$$\theta = \frac{\pi l}{\lambda}(\mu_a - \mu_b) \qquad (6.8)$$

where l is the path length, λ is the wavelength, and μ_a and μ_b are the refractive indices along the principal directions.

Light linearly polarized by a filter that then has its plane of polarization rotated through an angle θ after passing through a material before emerging through a second polarizing filter (Figure 6.3.2.2) has its intensity modulated. When the input and output polarizing filters are aligned, the output intensity variation is (e.g., Guenther, 1990)

$$I \alpha \sin^2 \theta \qquad (6.9)$$

Equation 6.8 indicates that the rotation angle θ varies inversely as the optical wavelength and any wavelength dependence of the refractive indices difference $(\mu_a - \mu_b)$. Consequently, different optical wavelengths are rotated through different angles θ. Hence, with incident white light, the "color" of the light emerging from a crossed polarizer depends upon the geometric path length and any other factors that might affect the refractive indices difference.

To make fundamental processes more transparent $(\mu_a - \mu_b)$, Equation 6.8 may be replaced by material properties that are specific to the category (birefringent, optically active, magneto-optic) to which the material belongs. Such expressions are considered in the following sections.

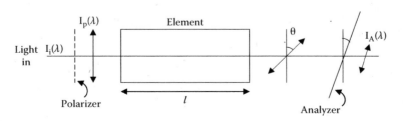

FIGURE 6.3.2.2
Optical addressing of an optically active element, 1 (From Jones, G.R., Li, G.D., and Spencer, J.W., Aspey, R.A., and Kong, M.G., *Opt. Commn.*, 145, 203–212, 1998. With permission.).

6.3.3 Polychromatic Birefringence

6.3.3.1 Basic Theory

The rotation of the plane of polarization of linearly polarized light θ propagating through a birefringent material has been described in Section 6.3.2 by Equation 6.8. It was indicated that as θ is wavelength dependent, different wavelengths are rotated through different angles. The net effect is that colored interference fringes are produced (rather than bright and dark ones obtained with monochromatic light). The intensity of the polychromatic light emerging from the material when the polarizing and analyzing filters are orthogonal is of the form given by Equation 6.7.

Extension to the case of polychromatic stress-induced birefringence yields the following equation (Murphy and Jones, 1992; Murphy and Jones, 1993)

$$I = \sum I_\lambda \sin^2(2\beta)\sin^2(\theta(\lambda)) \tag{6.10}$$

where β is the angle between the polarizing filter and the principal strain, I_λ is the optical intensity at a wavelength λ.

The anisotropy causing birefringence in photoelastic materials is produced by a strain ε. The refractive indices of the material parallel and orthogonal to the stain direction are related to the strain via the strain optic coefficient of the material k, i.e.,

$$(\mu_a - \mu_b) = \varepsilon K \tag{6.11}$$

Consequently, the light intensity emerging from the material varies with the magnitude of the strain.

6.3.3.2 Chromatic Processing

The detection of colored fringe changes as θ (Equation 6.8) is varied may be monitored chromatically using three nonorthogonal detectors. (The output from the detectors may be processed to yield chromatic parameters (H, L, S or x, y, z) so that the sequence of chromatic fringe changes may be tracked on a chromatic map. Figure 6.3.3.1 (Murphy and Jones, 1993) shows the locus of points corresponding to such a fringe sequence as calculated using Equation 6.2 with an ideal light source ($I = 1$) on an x:y chromatic map.

The results shown in Figure 6.3.3.1 are computer simulations obtained with Equations 3.4 and 3.5 (Chapter 3, Section 3.3.1).

$$x = R_0/(R_0 + G_0 + B_0) \tag{6.12}$$

$$y = G_0/(R_0 + G_0 + B_0) \tag{6.13}$$

These show that the locus for increasing values of θ is a convolute of variable radius about the achromatic point (1–2), Figure 6.3.3.1.

For CCD-based photoelastic monitoring with a broadband source (e.g., Ahmed, 1997) Equations 6.12 and 6.13 may be utilized. With optical fiber-based sensing it is more convenient to use a two-detector (distimulus) system

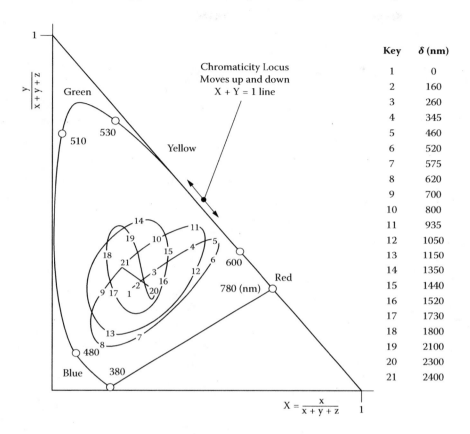

FIGURE 6.3.3.1
Locus of points on an *x-y* chromatic map corresponding to a sequence of polychromatic fringes (crossed polarizers, theoretical calculation, Equation 6.2, $x \equiv R$, $y \equiv G$, $z \equiv B$). (From Murphy, M.M. and Jones, G.R., *Pure Appl. Opt.*, 2, 1993. With permission.)

due to practical constraints of capturing the light emerging from the optical fiber. Consequently, Murphy and Jones (1993) have utilized algorithms based upon Equations 3.13 and 3.14 (Chapter 3, Section 3.3.2a), i.e.,

$$x = R_0/(R_0 + G_0) \qquad (6.14)$$

$$y = G_0/(R_0 + G_0) \qquad (6.15)$$

6.3.3.3   Chromatic Monitoring of Photoelasticity

Monitoring mechanical stress with a photoelastic material is an example of how chromatic methods can be used for quantifying changes produced by birefringence (Figure 6.3.3.2a); Murphy and Jones (1993) have shown typical changes in the spectrum of light propagated through a photoelastic material subjected to different levels of strains (0–1.8 × 10³ μm/m). Although there are

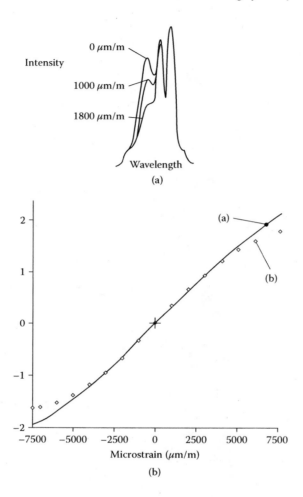

FIGURE 6.3.3.2
Chromatic photoelastic strain gauge (a) illustration of typical spectral changes (intensity vs. wavelength) produced by three different tensile strains (0, 10^3, 1.8×10^3 µm/m); (b) transfer characteristic (chromatic output: strain Continuous line (a)—theoretical curve; circles (b)—experimental points. (From Murphy, M.M. and Jones, G.R., *Pure Appl. Opt.*, 2, 1993. With permission.)

only limited spectral changes at the longer wavelengths, there are substantial changes distributed throughout the shorter wavelength range, making addressing with chromatic technique advantageous. The approach not only allows interfringe measurements to be made but also enables fringe orders to be distinguished (Kershaw et al., 1990).

Photoelastic sensors may be practically deployed either with optical fiber or with remote CCD addressing systems. Figure 6.3.3.2b shows the transfer characteristic of a chromatically addressed optical fiber sensor in the form of the ratio y/x (Equations 6.12 and 6.13) plotted against microstrain. Such sensors have considerably smaller operational areas than electronic ones ($\sim 5 \times 10^{-8}\,m^2$ compared to $\sim 5 \times 10^{-5}\,m^2$) (Murphy and Jones, 1993) and offer almost point strain measurement.

In the case of remote monitoring with a CCD camera (Ahmed et al., 1997), the photoelastic element is of greater dimensions than in the case of the optical fiber sensor. A 2-D image of a distributed, polychromatic fringe pattern on such an element used to measure pressure shows a complex distribution of stresses across the element (Figure 6.1.1). The 2-D fringe pattern may be divided into a series of concentric annular regions (Figure 6.1.1c, e.g., Ahmed et al., 1997) and mean values of the chromatic parameters *x, y* calculated for each annulus for a range of different strains (pressures). Typical results for two such annular regions (5, 10) show a complicated variation of each of the two chromatic parameters (Figure 6.3.3.3a; Ahmed et al., 1997) which are cyclic in nature. The corresponding *x-y* maps are complicated convolutes of the form shown on Figure 6.3.3.1 but differing significantly for the two (5, 10) annular regions. However by further processing of the chromatic parameter variations with a trained neural network, a pressure calibration curve with an RMS error of only 2% (Figure 6.3.3.3b) is achievable. This is an illustration of how distributed chromatic information can be used to provide acceptable accuracy monitoring even when highly localized values appear of limited use.

6.3.4   Polychromatic Rotary Activity

6.3.4.1   *Rotary Active Liquids*

For optically active fluids the rotation of the plane of polarization (Equation 6.8) becomes (Hart, 1999):

$$\theta = \frac{\pi.l.(u_a - u_b)}{\lambda} = \left[\alpha_\lambda^T\right].c.l \qquad (6.16)$$

where $[\alpha_\lambda^T]$ is the specific activity, *c* is the concentration of the rotary active material in solution (g/mL), and *l* the path length. A knowledge of θ allows the concentration of the optically active material to be determined provided its specific activity $[\alpha_\lambda^T]$ is known.

The specific activity $[\alpha_\lambda^T]$ varies with wavelength according to the equation (Crabbé, 1972):

$$[a_\lambda] = \frac{A}{\left(\lambda^2 - \lambda_c^2\right)} \qquad (6.17)$$

Where λ_c is the wavelength associated with the dominant molecular interaction in the material and *A* is a constant of proportionality depending upon the molecular weight of the optically active compound. Consequently, polarized light of different wavelengths will have their plane of polarization rotated by different amounts, the longer wavelengths being rotated more than the

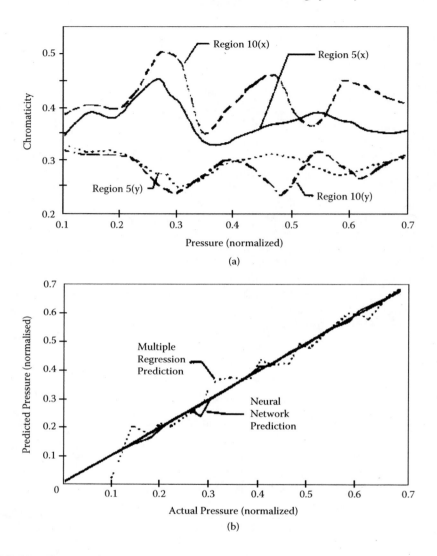

FIGURE 6.3.3.3
Polychromatic photoelastic membrane remotely addressed by a CCD camera for monitoring pressure: (a) variation of x,y chromatic coordinates with pressure averaged over each of two concentric annuli on the membrane (regions 5, 10); (b) neural network predicted pressure vs. actual pressure (solid line—neural network, dashed line—multiple regressions). (From Ahmed, S., Russell, P., Lisboa, P., and Jones, G.R., *IEE Proc.—Sci. Meas. Technol.*, 144, Fig. 6, 1997. With permission.)

shorter ones (rotary dispersion). Thus, if polychromatic—rather than mono-chromatic—light is used, the different rotation of the various wavelengths leads to the color of the light viewed through the analyzing filter varying with the angle of the analyzer is shown in Figure 6.3.4.1. The optical activity of the material can therefore be determined in practice without the need to

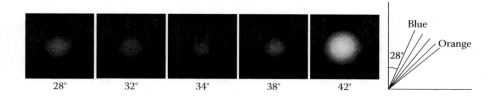

FIGURE 6.3.4.1
(See color insert following page 18). Chromatic images as a function of polarization angle (28–42°) for maltodextrine (Egan, 2006). (Sketch shows relative indications of the polarization for each image.)

rotate the analyzing filter but via the color of the emerging light, which can be quantified using chromatic techniques.

The approach can be used for addressing the following complex situations (Egan, 2006):

- The determination of the concentration, c, of an optically active compound (e.g., sucrose) in aqueous solution

- The determination of the concentrations of two different optically active compounds (e.g., sucrose and tartaric acid mixed in an aqueous solution)

- The determination of the concentration and molecular weight of optically active compounds during their formation (e.g., maltodextrine)

Examples of each of these conditions follow. In each case the light source used was a halogen lamp and chromatic detection achieved via a CCD camera.

6.3.4.1.1   Single Specy Concentration

Figure 6.3.4.2a,b,c show respectively the variation of each of the chromatic parameters H, L, and S with the inclination of the analyzing polarizing filter with sucrose concentration (0–0.15 g/mL) as the parameter. The results illustrate the highly complicated nature of the changes occurring in H, L, and S with the variations in the magnitude of each parameter being a substantial fraction of the full scale ($\Delta H \sim 200°/300°$, $\Delta L \sim 0.8/1$, $\Delta S \sim 0.65/1$).

To avoid the need to rotate the analyzer polarizer, a suitable angle needs to be chosen based upon the following criteria:

- There needs to be a reasonable change in each of the parameters over the concentration range to be monitored.

- It is preferable for the variation of each parameter with concentration to be monotonic to avoid unnecessary complications.

- The analyzer angle chosen should be in a range within which each chromatic parameter should not vary acutely with analyzer angle to avoid stringent alignment demands.

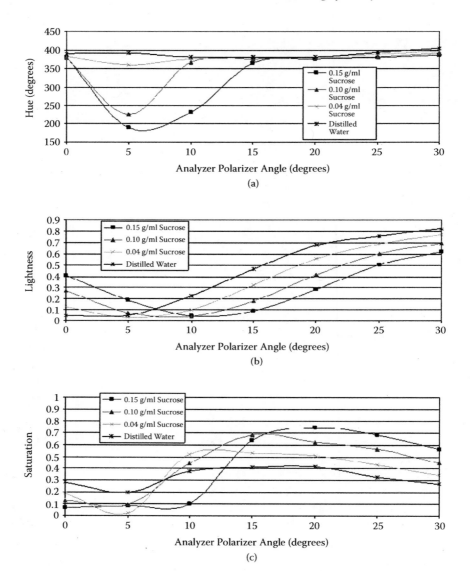

FIGURE 6.3.4.2
Variation of chromatic parameters *H*, *L*, and *S* for solutions of sucrose with different concentrations: (a) *H*-analyzer angle for different concentrations; (b) *L*-analyzer angle for different concentrations; (c) *S*-analyzer angle for different concentrations; (d) *H*, *L*, *S*-concentration for an analyzer angle of 30° (Egan, 2006).

Inspection of Figure 6.3.4.2a,b,c suggests that an angle of 30° would satisfy all three constraints. The variation of *H*, *L*, and *S* at this analyzer angle as a function of concentration is given in Figure 6.3.4.2d. The changes in values of each of the parameters (($\Delta H \sim 200°/360°$, $\Delta L \sim 0.2/1$, $\Delta S \sim 0.3/1$) is reasonable

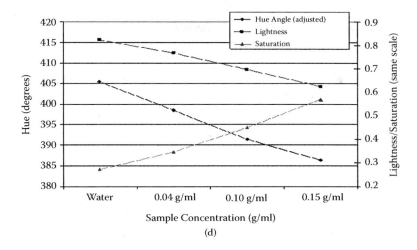

FIGURE 6.3.4.2
(Continued).

to provide satisfactory resolution for the concentration range. The variation of each of the parameters (H, L, and S) with concentration is approximately linear, so facilitating the determination of the concentration, c from the measured values (H, L, and S):

$$\begin{bmatrix} H \\ L \\ S \end{bmatrix} \cong \begin{bmatrix} a_h \\ a_l \\ a_s \end{bmatrix}.c + \begin{bmatrix} b_h \\ b_l \\ b_s \end{bmatrix} \tag{6.18}$$

A cross correlation between the values of each chromatic parameter measured at a given concentration can provide a cross-check for the reliability of the measurement. As a cross-check, Figure 6.3.4.3a shows an H-S diagram covering the 0–0.15 g/mL concentration range derived from the results given in Figure 6.3.4.2d. This shows a clear monotonic, almost linear variation of H with S. The departure of an experimental measurement from the curve would be indicative of the reliability of the measurement and the extent to which there could be intruding effects. The approximately linear nature of the curve facilitates estimates of such departures from the anticipated behavior.

This example has shown how, by careful use of the chromatic transformation, complicated outputs from a monitoring system can be simplified to conveniently yield values of the parameters being sought.

6.3.4.1.2   Concentration and Composition of a Two-Species Mixture

The chromatic approach has the potential for discriminating both the concentration and composition of a solution of two different optically active components. By suitable choice of inclination of the analyzing polarizing filter, using calibrations of H, L, and S similar to those shown in Figure 6.3.4.2a,b,c, a suitable

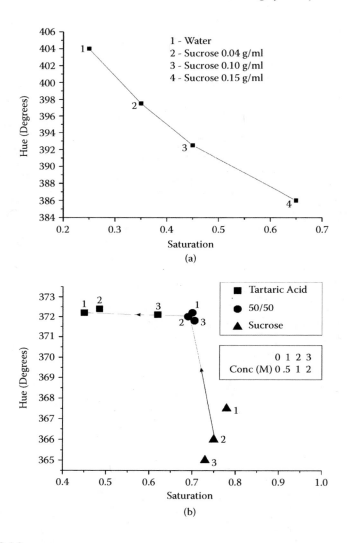

FIGURE 6.3.4.3

Examples of *H-S* chromatic maps for various optically active aqueous solutions: (a) single component (sucrose), various concentrations (0–0.15 g/mL), 30° analyzer angle; (b) two-component
(sucrose and L-tartaric acid), various concentrations and compositions (0–100%), 65° analyzer
angle (arrows—increasing L-tartaric concentration); (c) variable molecular weight and concentration (maltodextrine) 38° analyzer angle (arrows—increasing molecular weight (MW))
(Egan, 2006).

angle can be chosen to accommodate the range of conditions to be addressed.
A chromatic map using the most appropriate chromatic parameters may then
be chosen for distinguishing between the species and the concentrations.

   One such possible map is shown in Figure 6.3.4.3b, based upon the *H-S*
coordinates for an analyzer polarizer inclination of 65°. This map is for pure

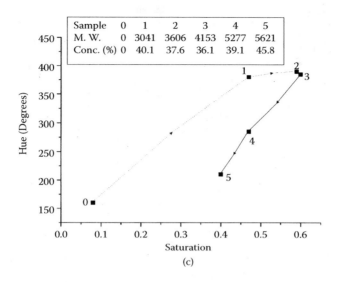

Sample	0	1	2	3	4	5
M. W.	0	3041	3606	4153	5277	5621
Conc. (%)	0	40.1	37.6	36.1	39.1	45.8

FIGURE 6.3.4.3
(Continued).

sucrose and L-tartaric acid solutions of different concentrations (0–2 M) and also a 50/50 mixture of sucrose and L-tartaric acid of various concentrations. The pure sucrose and L-tartaric acid solutions are discriminated by being in different regions of the map. Sucrose occurs at $0.74 < S < 0.78$, $365° < H < 368°$ and L-tartaric acid at $0.46 < S < 0.62$, $372° < H < 373°$. Sucrose concentration changes are dominated by changes in H, whereas L-tartaric acid concentrations are dominated by changes in S. The 50/50 mixture of the two components is distinguished from both single-component solutions in having distinctive H-S coordinates ($S \sim 0.7$ $H \sim 372°$). This particular analyzer angle is effective in discriminating the mixture from the pure components but not the concentrations; a separate analyzer angle can be chosen to emphasize the latter once the mixture is identified.

6.3.4.1.3   Concentration and Molecular Weight Discrimination

Optically active compounds can form long polymer chains (optically active molecules combined in long chains of various lengths), one example being maltodextrine (Hardy, 2007). As the chain length of the molecule grows so does the molecular weight, and its optical activity varies. This increase in molecular weight can be tracked using chromatic monitoring of the optical activity.

The procedure is to identify a suitable analyzer inclination by examining how the chromatic parameters H, L, and S vary with analyzer inclination for a range of maltodextrine molecular weights. The process is similar to that shown in Figure 6.3.4.2 for sucrose. An analyzer angle of 28° may be chosen as a suitable angle, which leads to the *H-S* chromatic map of Figure 6.3.4.3c.

This figure shows that the effect of the molecular weight and concentration variations can be discriminated via the nonmonotonic nature of the chromatic locus. An increase in the concentration is heavily associated with an increase in *S* (concentration increasing with order of points (3, 2) (4, 1, 5) in Figure 6.3.4.3c). An increase in molecular weight is more associated with an increase in *H* (points 1, 4 in Figure 6.3.4.3c have similar concentrations but different molecular weights). This is an illustration of the usefulness of chromatic mapping in assisting to discriminate trends in two properties (*MW*, concentration) via two chromatic parameters (*H, S*) when both the latter are coupled to each other but the former only weakly so.

6.3.4.1.4   Conclusions

The results presented as *H-S* chromatic maps in Figure 6.3.4.3a,b,c are examples of how different aspects of the aqueous solutions of optically active materials can be monitored chromatically in a versatile and yet economic manner. Not only do the maps define loci that yield specific information (e.g., concentration and composition) as stipulated *a priori* by the user, but they also enable unexpected departures from the intended operation to be identified.

6.3.4.2   Chromatically Addressed Thermochromic Elements

Thermochromic materials have optically active helical structures whose pitch varies significantly with temperature (Elser and Ennulat, 1976) (Figure 6.3.2.1). Consequently, the specific activity $[\alpha_\lambda T]$ (Equation 6.17) and phase difference between two circularly polarized components of a plane polarized wave propagating through the material (Equation 6.6; Figure 6.3.1.1) varies. This leads to a change in the plane of polarization of the emerging light. More significantly, the changes in this plane also vary with optical wavelength. As a result, the chromaticity of a broadband beam emerging from such a material varies with temperature.

An example of such thermochromic chromatic changes is given in Figure 6.3.4.4. Each frame of this figure shows a horizontal row of five thermochromic elements, each having a 5°C temperature range but with the ranges covering different temperature levels (Deakin et al., 2005). As the temperature increases the chromaticity of an element (e.g., element 2, Figure 6.3.4.4a,b) changes. In a different temperature range, the chromaticity of a second element (element 3, Figure 6.3.4.4b,c) becomes more responsive to the temperature.

Figure 6.3.4.5a shows the variation of the output of each of the *R, G, B* processors of a chromatic monitor based upon a CCTV camera (Ahmed et al., 1997). Each processor output is nonmonotonic, thus producing ambiguities in a temperature value determined from a single processor. However, by taking the output of all three processors into account, such ambiguity can be removed and an accurate measurement of temperature obtained. Ahmed et al. (1997) showed that such data could be used to train a neural network, which could

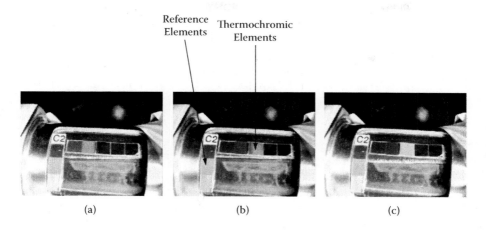

Reference Elements   Thermochromic Elements

(a)   (b)   (c)

FIGURE 6.3.4.4
(See color insert following page 18). Images of thermochromic elements at different temperatures. (Temperature increasing in the order (a), (b), (c) as chromaticity of elements changes in the order 2, 3 from the left.)

yield a temperature from a previously unseen data set to an accuracy of 0.5% over a temperature range from 25°C to 80°C (Figure 6.3.4.5b).

6.3.5   Chromatically Addressed Magneto-Optic Elements

The phase difference between the differently polarized waves propagating in a magneto-optic material is given by Equation 6.8. The refractive index term may be rewritten in terms of a material specific constant $V(\lambda)$ (the Verdet constant) and the magnetic field strength of density B to which it is exposed (Li et al., 1999b), i.e.,

$$\theta = \frac{\pi l}{\lambda}(\mu_a - \mu_b) = V(\lambda)Bl \qquad (6.19)$$

where l is the path length of the light and $V(\lambda)$ is dependent upon the optical wavelength.

If a magneto-optic element is placed in between two linear polarizing filters (e.g., Figure 6.3.2.2) that polarize in the same direction, the intensity of the emergent light varies with θ according to Equation 6.9 in Section 6.2.

The magneto-optic effect may be used for monitoring electric currents via the magnetic field that is produced in proportion to the current (Jones et al., 1998; Li et al., 1999a, 1999b). There are three ways in which such a magneto-optic transducer can be realized chromatically as a current sensor:

1. Wavelength-dependent Verdet constant (Jones et al., 1998) with which chromatic modulation of the polarized polychromatic light is produced solely by the wavelength dependence of the Verdet constant. Figure 6.3.5.1 shows the shift in the cyclical polychromatic intensity as

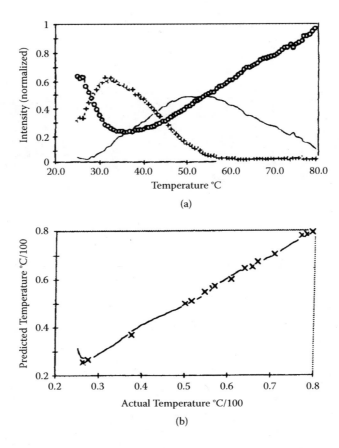

(a)

(b)

FIGURE 6.3.4.5
Chromatic processing of a thermochromic element for temperature monitoring: (a) variation of
normalized *R, G, B* processor outputs with temperature; (b) predicted vs. measured temperature from neural network trained system. (From Ahmed, S., Russell, P., Lisboa, P., and Jones,
G.R., *IEE Proc.—Sci. Meas. Technol.*, 144, 1997. With permission.)

a function of wavelength produced by a *B* field affecting the Verdet constant. The responsivities of two nonorthogonal chromatic detectors R_A,
R_B are also shown in Figure 6.3.5.1a to illustrate how their responses are
affected by the shift in polarization (Jones et al. 1998; Li et al., 1999a).

2. Referenced magneto-optic rotation (Jones et al., 1998) with which
   the intensity change of the polarized shorter wavelengths (visible)
   light caused by the magneto-optic rotation is compared with the
   intensity of the infrared light from the same source that is unpolarized because it lies at wavelengths beyond those affected by the
   polarizing filters. The configuration of the transducer is similar
   to that shown in Figure 6.3.2.2. Figure 6.3.5.1b (Jones et al., 1998)
   shows the responsivities of two chromatic detectors overlapping

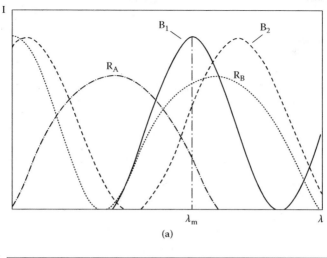

(a)

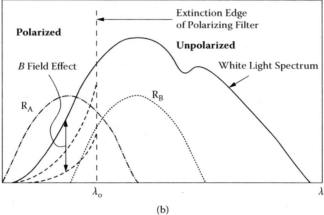

(b)

FIGURE 6.3.5.1
Wavelength dependent magneto-optic transducer: (a) change in the cyclical light intensity caused by a B field change ($B_1 -> B_2$) superimposed upon two nonorthogonal detector responses R_A, R_B ($V(\lambda) \propto \lambda$); (b) spectral changes produced by the magneto-optic rotation of polarized light for wavelengths shorter than the extinction edge of the polarizing filter superimposed upon two nonorthogonal filters R_A, R_B. (No changes in the unpolarized light at wavelengths longer than the extinction edge.) (From Jones, G.R., Li, G.D., and Spencer, J.W., Aspey, R.A., and Kong, M.G., *Opt. Commn.*, 145, 1998. With permission.)

the visible wavelengths below the extinction edge of the polarizing filter and the unpolarized wavelengths above the extinction edge.

3. Wavelength-dependent rotary power element (Li et al., 1999, 1999a; Jones et al., 1998) with which chromatic modulation of polarized

polychro matic light from a magneto-optic element, which is only weakly wavelength-dependent, is wavelength encoded by an element that has a natural rotary power (Section 6.3.2). A schematic of the transducer is given in Figure 6.3.5.2a. This shows the rotary

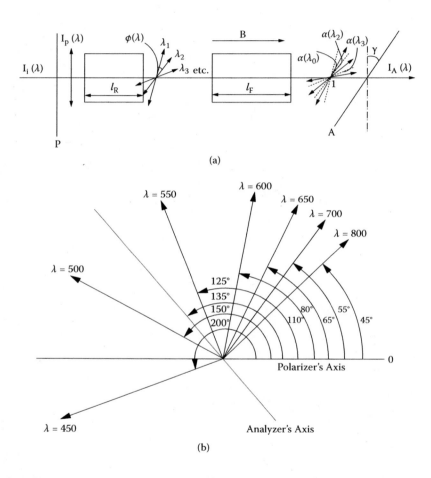

FIGURE 6.3.5.2
Combined magneto-optic and wavelength dependent rotary elements transducer: (a) schematic of transducer system (From Jones, G.R., Li, G.D., and Spencer, J.W., Aspey, R.A., and Kong, M.G., *Opt. Commn.*, 145, 1998. With permission.); (b) rotation of the plane of polarization at different wavelengths produced by a 4.25-mm-thick quartz plate (From Li, G.D., Aspey, R.A., and Jones, G.R., 1999a, *Opt. Commn.*, 162, 1999. With permission.) (c) spectral shift produced by a change in B field ($B = 0 \ddagger B_1$) for the combined magneto-optic/rotary power transducer. (From Jones, G.R., Li, G.D., and Spencer, J.W., Aspey, R.A., and Kong, M.G., *Opt. Commn.*, 145, 1998. With permission.)

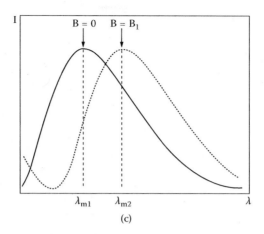

(c)

FIGURE 6.3.5.2
(Continued).

power unit producing wavelength encoding at various polariza-
tion angles $\varphi(\lambda)$, details of the encoding of wavelength as a function
of angle being shown in Figure 6.3.5.2b for a quartz encoder. The
wavelength encoded polarization inclinations are then fed through
the magneto-optic element that rotates all wavelengths similarly.
The output wavelengths emerging from the analyzing polarizer
depend upon the additional degree of rotation produced by the B
field to which the magneto-optic element is subjected. Figure 6.3.5.2c
shows the kind of spectral shift produced by such a unit.

Typical characteristics relating a current-produced B field to the dominant wavelength as determined with a two nonorthogonal detector system
is shown in Figure 6.3.5.3. Results are given for the wavelength-dependent
Verdet constant (i) and the wavelength-dependent rotary power element (iii),
which shows the improved sensitivity obtained with the rotary power device.
A comparison of the performance of the three chromatically based magneto-
optic methods for measuring electric current (i), (ii), and (iii) is given in
Table 6.1 (Li et al., 1999a). This comparison shows that white-light LED-based
systems have superior performances than tungsten halogen lamp systems
and that the systems employing a wavelength-dependent rotary power element has a greater dynamic range ($\nabla\lambda$) and higher sensitivity and resolution
than a wavelength-dependent Verdet constant element.

TABLE 6.1

Comparison of Various Chromatically Based Magneto-Optic Systems for Monitoring Electric Currents

System Type		$V\lambda$(nm)[4]	Sensitivity nmA^{-1}	Resolution A/$\sqrt{Hz}$
			Tungsten-Halogen	
I.R. referenced system[1]		−5	0.025	0.01
Wavelength-dependent rotary power element[2]	BSO	9.5	0.19	0.0012
			"White" LED	
Wavelength-dependent Verdet constant[3]		15	0.4	0.00055
Wavelength-dependent rotary power element	Quartz	26	0.065	0.00034

[1] Performance limited by temperature dependence of extinction edge of polarizing filter. (Not relevant to LED with no infrared.)
[2] Optical attenuation high in visible spectrum, part of infrared filtered.
[3] With tungsten-halogen source, changes in visible spectrum small compared with unpolarized infrared spectrum.
[4] Minimum measurable wavelength change ~0.01 nm.

Source: Adapted from Li, G.D., Aspey, R.A., and Jones, G.R., *Opt. Commn.*, 162, 1999a. With Permission.

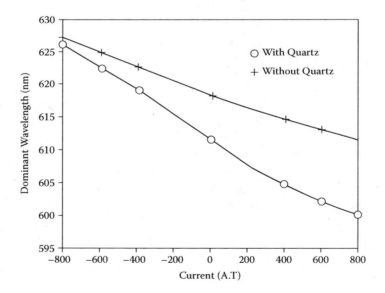

FIGURE 6.3.5.3
Dominant wavelength as a function of current in amp-turns with and without a quartz rotary encoder. (From Li, G.D., Aspey, R.A., and Jones, G.R., 1999a, *Opt. Commn.*, 162, 1999. With permission.)

6.4 Summary

The complex signals produced from broadband optical interferometry and polarimetry may be addressed with chromatic techniques for extracting relevant information in a robust and efficient manner.

Both spectral- and phase-domain interferometric signals can be addressed. The former utilizes chromatic processors deployed in the optical wavelength domain spanning the spectral range of interest and enabling interfringe distances and fringe order to be determined. The latter utilizes chromatic processors deployed in the time domain in a time-stepping mode to improve the spatial discrimination of interference patterns.

Broadband birefringence in anisotropic materials may be addressed with chromatic methods via either optical-fiber transmission or remotely by CCD cameras for monitoring externally induced conditions such as pressure. It may be simultaneously indicated whether extraneous effects, such as temperature variations, are occurring.

Broadband optical polarization effects in optically active materials may be observed with chromatic methods deployed in the wavelength domain. The concentration and composition of binary mixtures of optically active components in a solution may be determined from changes in the chromatic parameters H, L, S. The molecular weight and concentration evolution during the formation of long chains of optically active polymers can be tracked. Temperature may be monitored using thermochromic elements addressed via optical fibers or remotely with CCD cameras.

Wavelength-domain chromatic techniques have been deployed in conjunction with magneto-optic materials for measuring electric currents. A combination of a magneto-optic element in series with an optically active element can improve both the resolution and range of such electric current sensors.

These represent examples of only a few of the uses of chromatically deployed interferometric and polarimetric methods. A further example of the application of phase domain interferometry is described for monitoring biological tissue in Chapter 8.

References

Ahmed, S.U., 1998. Intelligent Remote Chromatic Processing, Ph.D. thesis, University of Liverpool.

Ahmed, S.U., Russell, P., Lisboa, P., and Jones, G.R., 1997. Parameter monitoring using neural-network-processed chromaticity, *IEE Proc.–Sci. Meas. Technol.*, 144, 257–262.

Crabbé, P., 1972. *ORD and CD in Chemistry and Biochemistry*, Academic Press, New York.

Deakin, A.G., Rallis, I., Zhang, J., Spencer, J.W., and Jones, G.R., 2005. Towards holistic chromatic intelligent monitoring of complex systems, *Proceedings of the Complex Systems, Monitoring Session, International Complexity, Science and Society Conference* (Liverpool), pp. 16–23.

Ditchburn, R.W., 1958. *Light*, Blackie & Sons, London.

Egan, C.A., 2006. Monitoring of optically active chemicals using chromatic modulation techniques, Ph.D. thesis, University of Liverpool.

Elser, W. and Ennulat, R.D., 1976. in *Advances in Liquid Crystals*, Vol. 2, Brown, G.H., Ed., Academic Press, New York, pp. 73–172.

Grattan, K.T.V. and Meggitt, B.T., 1995. *Optical Fiber Sensor Technology*, Chapman & Hall, London.

Guenther, R.D. (1990). *Modern Optics*, John Wiley, New York.

Hardy, J.K., Retrieved 2007. *The Chemical Database*, The University of Akron, http://ull.chemistry.uakron.edu/erd/chemicals/13000/12238.html.

Hart, H., 1999. *Organic Chemistry: A Short Course*, Houghton Mifflin, Boston.

Jones, G.R., 2001. Optical fibre-based monitoring of high voltage power equipment, in *High Voltage Engineering and Testing*, Ryan, H.M., Ed., IEE, London, pp. 601–634.

Jones, G.R., Russell, P., and Khandaker, I., 1994. Chromatic interferometry for an intelligent plasma processing system, *Meas. Sci. Technol.*, 5, 639–647.

Jones, G.R., Li, G.D., and Spencer, J.W., Aspey, R.A., and Kong, M.G., 1998. Faraday current sensing employing chromatic modulation. *Opt. Commun.*, 145, 203–212.

Jones, G.R., Jones, R.E., and Jones, R., 2000. Multimode optical fiber sensors, in *Optical Fiber Sensor Technology*, Grattan, K.T.V. and Meggitt, B.T., Eds., Kluwer Academic, Dordrecht, The Netherlands, pp. 1–77.

Kershaw, D., Pridham, B., Jones, G.R. et al. 1990. Image processing techniques in photoelasticity, *Proceedings of the Applied Optics and Opto Electronic Conference* (Institute of Physics, Nottingham, UK), pp. 208–209.

Li, G.D., Aspey, R.A., and Jones, G.R., 1999a. White LED-based Faraday current sensor using a quartz wavelength encoder, *Opt. Commn.*, 162, 44–48.

Li, G.D., Aspey, R.A., and Kong, M. G., Gibson, J.R., and Jones, G.R., 1999b. Elliptical polarization effects in a chromatically addressed Faraday current sensor, *Meas. Sci. Technol.*, 10, 25–31.

Meggitt, B.T., 1995. Fiber optic white light interferometric sensors, in *Optical Fiber Sensor Technology*, Grattan, K.T.V. and Meggitt, B.T., Eds., Chapman & Hall, London, pp. 269–312.

Messent, D.N., Singh, P.T., Humphries, J.E., et al., 1997. Optical fibre measurements of contact stalk temperature in an SF6 circuit-breaker following fault current arcing, *Proceedings of the XII International Conference on Gas Discharges and Their Applications* (Greifswald).

Murphy, M.M. and Jones, G.R., 1992. Polychromatic birefringence sensing for optical fibre monitoring of surface strain, *Sensors and Actuators. A, Physical*, 32, 691–695.

Murphy, M.M. and Jones, G.R., 1993. An extrinsic integrated optical fibre strain sensor, *Pure Appl. Opt.*, 2, 33–49.

Russell, C.D., Bullough, T.J., Jones, G.R., Krasner, N., and Bamford, K.J., 2003. (2004) Application of chromatic analysis for resolution improvement in optical coherence tomography (OCT), *Proc. SPIE* Vol. 5486 p. 123–128.

Russell, C.D., Deakin, A.G., and Jones, G.R., 2005. Chromatic analysis for signal processing in optical coherence tomography, *Proceedings of the Complex Systems Monitoring Session, International Complexity., Science and Society Conference* (Liverpool), pp. 35–39.

Tilley, R.J.D., 2000. *Colour and the Optical Properties of Materials*, John Wiley, New York.

7

Particulates Monitoring with Polychromatic Light

G.R. Jones and Y.R. Kolupula

CONTENTS

7.1 Introduction

The monitoring of micron-sized particles is of increasing importance in industry and for environmental and health studies (Noble and Prather, 1998; Holgate et al., 1999). Methods for measuring such particulates extend from precision methods made under laboratory conditions (e.g., laser scattering (Johnson and Gabriel (1981)), high-resolution microscopy, etc. (Ross (2005), to more robust methods used under real-world conditions (e.g., particle microweighing) (Meyer et al. (2000)); "black-smoke" optics (Hitzenberger et al. (1999)). Chromatic techniques have the potential for bridging the gap between these methods and particularly for addressing situations that may be complicated by interfering effects.

One example of the deployment of particle-related chromatic monitoring has been given in Chapter 4, Section 4.2.2.4, for monitoring the heating of particulates entrained into a plasma jet. Other examples are for the monitoring of the concentration and size of particles using polychromatic light scattered from the particles (e.g., Mie scattering; Mie, 1908; Kerker, 1969) or light scattered/absorbed, etc., by a combination of a particle filter and the particles accumulated on the filter.

Although Mie-type scattering is theoretically well founded, the particle–filter combination is arguably a more realistic representation of the complicated conditions encountered outside the laboratory.

The application of the chromatic approach for particle monitoring under the complexity of real-world conditions is described.

7.2 MIE-Scattered Polychromatic Light

7.2.1 Wavelength Dependence of MIE Scattering

The scattering of light by an assembly of suspended micron-sized particles (Mie, 1908; Kerker, 1969) leads to a change in the intensity of scattered light, *I*, from a length dx of such an assembly given by

$$I/I_o = u \cdot dx \tag{7.1}$$

where I_o is the incident light intensity, and u is the Mie extinction coefficient given by

$$U = \int_{r1}^{r_2} NKr^2\, dr \tag{7.2}$$

where N is the number of particles per unit volume; r, the particle radius; K, the extinction cross-section, which takes account of the scattering effect being greater than simply the physical radius, r. K varies with optical wavelength as shown on Figure 7.1 (e.g., Meyer-Arendt, 1995).

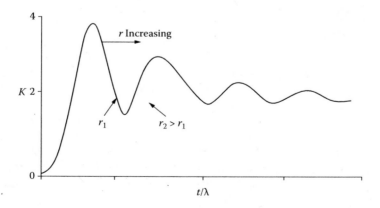

FIGURE 7.1

Extinction coefficient K as a function of wavelength with particle radius as parameter. (From Meyer-Arendt, J. R., (1995). *Introduction to Classical and Modern Optics*. Prentice Hall, Upper Saddle River, NJ. With permission.)

Equations 7.1 and 7.2 show that the intensity of light scattered depends upon the concentration (N) and size (r) of the scattering particles. Moreover, because the extinction cross section, K, varies with both the wavelength of the light and the particle radius (Figure 7.1), the intensity of light scattered at various wavelengths is different. Consequently, the signal strengths produced by detectors having nonorthogonal responsivities in the wavelength domain (Appendix 1.I) are different. Therefore, the values of the chromatic parameters of the scattered light depend upon the concentration and size of the particles, provided the latter are sufficiently small (less than tens of microns; Kerker, 1969).

The angular variation of the Mie-scattered light is also wavelength dependent (Kerker, 1969), whereby longer wavelengths are preferentially scattered in a forward direction, whereas shorter wavelengths are scattered sideways.

7.2.2 Chromaticity of MIE-Scattered Light

The wavelength dependence of Mie-scattered light is illustrated by the images shown in Figure 7.2a,b. These correspond to light transmitted forward through water. Figure 7.2a is for water without particles; Figure 7.2b is for a suspension of 10-μm particles in the water. The images show that the color of the forward transmitted light is more yellow in the presence of the 10-μm particles, consistent with the preferential forward scattering of the longer-wavelength components of the incident white light.

Light transmitted through such suspensions of particles may be addressed by chromatic processors whose outputs are transformed into chromatic H, L, S parameters. The variation of H, L, S with particle concentrations in the range 0–64 g/m^3 and for three different particle sizes (1,3,9 μm) are shown in Figure 7.3.

(a) (b)

FIGURE 7.2
(See color insert following page 18). Images of forward-scattered light from a water medium: (a) no particulates, (b) 10-μm particulates.

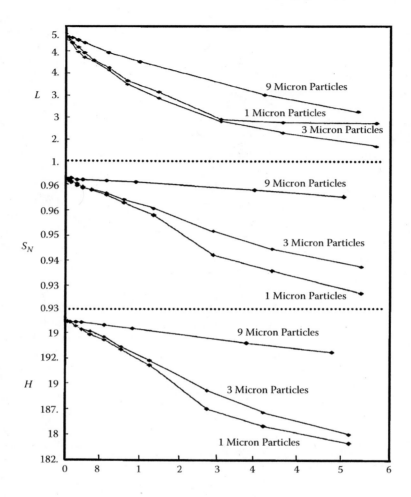

FIGURE 7.3
Variation of chromatic parameters L, S_N, H with particle concentration and with particle size as parameter for Mie-scattered light (particles suspended in water).

The results show that each of the chromatic parameters *H, L, S* varies with both particle concentration and size monotonically. The reduction in *H* with increasing particle concentration is consistent with a shift toward longer wavelengths of Mie scattering. The reduction in *L* is consistent with shorter wavelengths being scattered sideways and so reducing the forward transmitted light intensity. The *S* parameter is given as $S_N = 1 - S$ and shows the smallest relative changes of the three chromatic parameters. Consequently, the concentration and size of the particulates can be determined in principle by cross correlating the values of each of the chromatic parameters *H, L, S*.

Typical changes in *H, L, S* due to Mie scattering are within the ranges dL/L ~ 2.2/5.2, dH/H ~ 11/196, dS/S ~ 0.03/ 0.97 (Figure 7.3).

The chromatic approach, therefore, has the potential for monitoring the relatively ideal conditions of Mie scattering with only a single size of particle present, with well-dispersed particles in a transparent medium.

7.2.3 MIE Scattering Under Less Ideal Conditions

When high-current electric arcs are transiently formed in SF6 gas in circuit breakers, particles of about 10-μm dimensions are copiously formed and subsequently settle (Isaac et al., 1999). Figure 7.4 shows a sequence of images of white light being chromatically modified during the production of such particulates. It shows the color of white light progressively changing to an increasingly orange color (predominance of longer wavelengths) as the particle concentration increases, consistent with Mie scattering.

The chromaticity of the light scattered in this manner can be determined in its simplest form with only two optoelectronic detectors (distimulus) having nonorthogonal responsivities to yield the dominant wavelength (Chapter 3, Section 3.3.2(a)). Such simplified detection is advantageous when used with optical-fiber transmission to address electrically and chemically aggressive environments such as those of a high-voltage circuit breaker (Isaac et al., 1999). The time variation of the dominant wavelength measured from the formation of particulates during the sequential operation of a number of electric arcs in such a circuit breaker is shown in Figure 7.5a. The dominant wavelength increases rapidly following arcing, indicating copious particle formation, before decaying slowly as the particles settle. The dominant

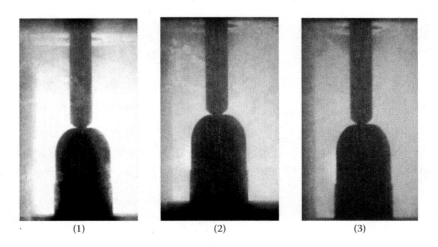

(1)   (2)   (3)

FIGURE 7.4
(See color insert following page 18). Images of forward-scattered light from particles formed by electric arcs in a SF6 high-voltage circuit breaker.

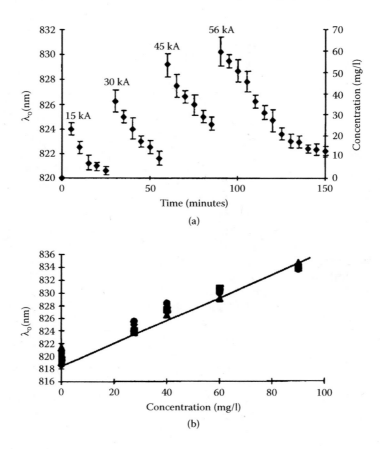

FIGURE 7.5
Particle formation by high-current arcs in SF6 circuit breakers: (a) dominant wavelength and particle concentration as a function of time, (b) dominant wavelength as a function of particle concentration. (From Isaac, L. T., Jones, G. R., Humphries, J. E., Spencer, J. W., and Hall, W. B. (1999). Monitoring particle concentrations produced by arcing in SF6 circuit breakers using a chromatic modulation probe. *IEE Proc. Sci. Meas. Technol.* 146, No. 4, pp. 199–204. With permission.)

wavelength varies in the range 820–830 nm with arc-sustaining currents of 15–56 kA over a period of 150 min. The high values of the dominant wavelength (820–830 nm) are due to the responsivities of the photodetectors extending into the infrared wavelength region.

Because the particles in this case are produced under controlled conditions in a reproducible manner, the dominant wavelength can be calibrated (Figure 7.5b) to give total particle mass concentration (Isaac et al., 1999). The measured dominant wavelength changes yield particle concentration changes in the range 0–60 mg/L and enable the growth, settling, and accumulation of particles to be quantified with respect to the circuit breaker operating conditions.

Inspection of quantities of samples of particles after a test indicated them to be of a slightly yellow color in reflected light, which could in principle affect the Mie-scattered light. However, the chromatic technique is capable via calibration of accommodating such possible aberrations.

7.3 Complex Scattering and Absorption

7.3.1 Particle Accumulation on Filters

Monitoring low concentrations of particulates can be assisted by accumulating the particles from a fluid flow onto a micropore filter, as is undertaken with TEOM microweighing systems (Meyer et al., 2000). Filter-accumulated particulates can also be addressed optically by transmitting polychromatic light through the particle-filter combination and monitoring the transmitted light chromatically. However the interaction of not only the particulates with the light but also the micropores of the filter (i.e., additional scattering and absorption) must be accommodated.

7.3.2 Practical Filter-Based System

A practical form of a chromatically addressed micropore system for monitoring 2.5–10-μm airborne particles is shown on Figure 7.6a. Airborne particles are drawn into a sealed unit via a filter that removes particles of size greater than 10 μm before being accumulated on a 2.5–10-μm microbore filter. The microbore filter is addressed with polychromatic light from a tungsten halogen lamp (or LEDs), which is monitored chromatically after transmission through the filter via a CCD array camera. The particle concentration range is adjustable by varying the rate at which the particle-bearing air is drawn into the unit and by varying the optical source voltage. The output from the CCD array provides the R, G, B values, which are transformed into the chromatic parameters H, L, S.

Examples of images of the microbore filter obtained with the CCD array are shown in Figure 7.6b. The filter is divided into three parts: the particle-gathering half and two reference quarters without particles, one representing the maximum output and the other the minimum output. The sequence of images shows how the chromaticity of the particle-gathering sector varies as particles are accumulated.

The chromatic stability of the system is indicated by the random variations of $H, S,$ and L being small: 2, 0.005 and 0.002°, respectively (Figure 7.7a,b,c). The variation between two sets of pixels is also small as shown by the two sets of points on each graph. The results are for a 24-h period during which no particulates were accumulated. The initial drift ($dH = 4°, dS = 0.015, dL = 0.015$) is due to the tungsten halogen source stabilizing immediately after switch on, and can be taken into account via the reference channels.

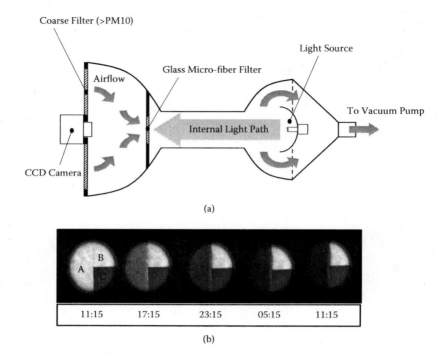

(a)

(b)

FIGURE 7.6
CCD-based particle accumulation system. (a) system schematic, (b) images of particle filter.
(From Reichelt, T. E., Aceves-Fernandez, M. A., Kolupula, Y. R. et al. (2006). Chromatic Modula-
tion monitoring of airborne particulates. *Meas. Sci. Technol.* 17, pp. 675–683. With permission.)

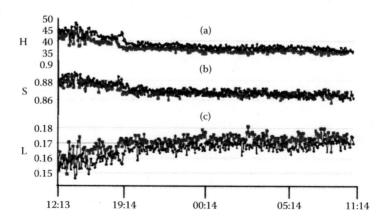

FIGURE 7.7
Time variation of chromatic parameters with no particle accumulation for two different sets of
CCD pixelss. (a) *H*, (b) *S*, (c) *L*. (From Reichelt, T. E., Aceves-Fernandez, M. A., Kolupula, Y. R.
et al. (2006). Chromatic Modulation monitoring of airborne particulates. *Meas. Sci. Technol.* 17,
pp. 675–683. With permission.)

7.3.3   Chromatic-L-Based Monitoring

The simplest chromatic approach is based upon tracking the time variation of L (Equation 1.3, Chapter 1) as 10-μm particles are accumulated on an illuminated particle filter (Figure 7.6). The lightness L of the signal sector (e.g., Figure 7.6b) is reduced compared with the reference levels as particles are accumulated. The time variation of lightness during particle accumulation is shown in Figure 7.8a (Reichelt et al., 2006).

The L values are calibrated against particle concentration with a series of microweighed filters carrying known concentrations of 10-μm (PM10) particles to produce calibration curves as shown in Figure 7.8b. A calibration curve for each of several, different lamp voltages (Reichelt et al., 2006) are obtained so that any variation in the signal from the reference sector of the filter can be used to identify the appropriate calibration curve. Application of the calibration graphs to the L results shown in Figure 7.8a leads to the time variation of accumulated particle concentrations, also shown in Figure 7.8a.

The chromatic L monitoring has significant differences from "black-smoke" methods (Hitzenberger et al., 1999) in that it incorporates online referencing using the two reference quarters of the filter, and the L values can be cross correlated with the chromatic S and H values to ensure correct system operation.

7.3.4   H-S Chromatic Parameters

The variation in chromatic L values with time resulting from particle accumulation (Section 7.3.3) may be accompanied by corresponding variations in H, and S. Some typical time variations for H, S, and L during particle accumulation are given in Figure 7.9a. During this test there was a deliberate change in the color temperature of the tungsten halogen light source by varying the lamp voltage from 3.5 to 6 V. This produces significant changes in all three chromatic parameters H, L, S. Thus, spurious changes in illumination can be indicated by cross correlating with the H and S values.

This may be shown by displaying H and L on a H-L polar diagram (Figure 7.9b). For a source color temperature corresponding to a 3.5-V lamp voltage, the major chromatic change is in the L parameter with H remaining fairly constant until $H \sim 0.3$. However, for a color temperature corresponding to 4.0 V, the chromatic parameter, which initially shows the dominant variation, is H ($30 < H < 60$) not L.

The 30° change in H with a 4-V lamp voltage is greater than the values of 3–11° expected from Mie-scattered light (Figure 7.3). This implies that the H shift is not simply due to Mie scattering but associated with the optical interaction with the microbore filter. The zero particle points on the 3.5- and 4.0-V characteristics being 30° apart indicates that the H difference is associated with the properties of the microbore filter. The accumulation of particles gradually reduces this difference and indicates the complex behavior of the system.

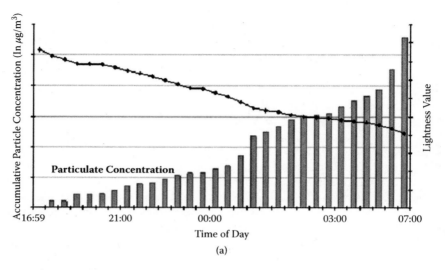

(a)

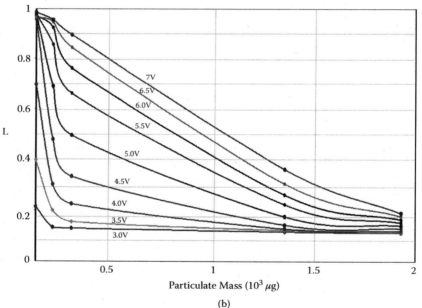

(b)

FIGURE 7.8
Chromatic L and particle concentrations. (a) L and concentration versus time (lamp voltage = 3.5 V), (b) L as a function of particle concentration for different lamp voltages (3.0–7.0 V). (From Reichelt, T. E., Aceves-Fernandez, M. A., Kolupula, Y. R. et al. (2006). Chromatic Modulation monitoring of airborne particulates. *Meas. Sci. Technol.* 17, pp. 675–683. With permission.)

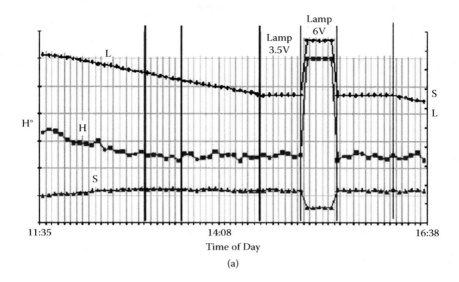

(a)

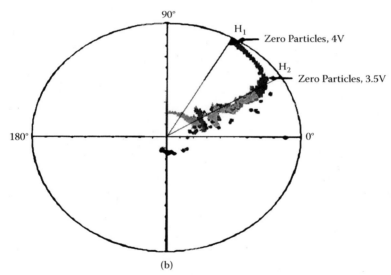

(b)

FIGURE 7.9
Variations of chromatic parameters with particle concentration and lamp voltages. (a) time variation of *H, L, S*, (b) *H-L* polar diagram for particle collection with two different lamp voltages (3.5, 4.0 V). (From Reichelt, T. E., Aceves-Fernandez, M. A., Kolupula, Y. R. et al. (2006). Chromatic Modulation monitoring of airborne particulates. *Meas. Sci. Technol.* 17, pp. 675–683. With permission.)

7.4 Summary

The Mie scattering of polychromatic light by micron-sized particles produces spectral changes that are chromatically detectable in terms of the parameters H, L, S. Under ideal conditions the concentration and size of the particles can, in principle, be determined by cross correlation between the H, L, S values.

In practice the Mie-scattered signals may be distorted or masked by other chromatic effects such as wavelength-dependent absorption, for example, when particles are accumulated on particle filters. However, the chromatic approach can still be utilized for particle monitoring under such conditions, using empirically determined calibration curves. Correct operation of the monitoring system can be checked via cross correlation between the various chromatic parameters.

References

Hitzenberger, R., Jennings, S. G., Larson, S. M., Dislner, A., Cachier, H., Galamos, Z., Ronc, A., and Spain, T. G. (1999). Intercomparison of measurement methods for black carbon aerosols. *Atmos. Environ.* 933, pp. 2823–2833.

Holgate, S., Samet, J., Koren, H., and Maynard, R. (1999). *Air Pollution and Health*. Academic Press, New York.

Isaac, L. T., Jones, G. R., Humphries, J. E., Spencer, J. W., and Hall, W. B. (1999). Monitoring particle concentrations produced by arcing in SF6 circuit breakers using a chromatic modulation probe. *IEE Proc. Sci. Meas. Technol.* 146, No. 4 pp. 199–204.

Johnson, C. S. and Gabriel, D. A. (1981). *Laser light scattering*. Dover, New York.

Kerker, M. (1969). *The Scattering of Light and Other Electromagnetic Radiation*. Academic Press, New York.

Meyer-Arendt, J. R. (1995). *Introduction to Classical and Modern Optics*. Prentice Hall, Upper Saddle River, NJ.

Meyer, M. B., Patashnick, H., Ambs, J. L., and Rupprecht, E. (2000). Development of a simple equilibration system for the TEOM continuous PM monitor. *J. Air and Waste Management Association* (August).

Mie, G. (1908). Beitrage zur optik truber medien, speriell koloidaler metallosungen. *Ann. Phys.* 4, 925 pp. 377–445.

Noble, C. and Prather, K. (1998). Air pollution: The role of particles. *Physics World* (January).

Reichelt, T. E., Acevez Fernandez, M. A., Kolupula, Y. R. et al. (2006). Chromatic Modulation monitoring of airborne particulates. *Meas. Sci. Technol.* 17, pp. 675–683.

Ross, I. A. (2005). Chromatic monitoring of particulates arising from air pollution. M.Phil Thesis, University of Liverpool.

8

Chromatic Monitoring of Biological Tissues and Fluids

E. Borisova, G.R. Jones, P. Pavlova, and C.D. Russell

CONTENTS

8.1   Introduction

Optical techniques provide a useful and powerful means for monitoring biological tissue and fluids such as human skin and blood because they are convenient to apply and often involve only a minimal invasion of the tissue or fluid when used *in vivo*. However, such tissues and fluids provide stern challenges for optical monitoring because of their inhomogeneous structures, complex chemistry, and variable nature. For example, cutaneous tissue consists of skin layers—epidermis and derma—and fatty tissue, muscle, cartilage, blood-filled capillaries, sweat glands, sebaceous glands, hair follicles, etc. A simplified diagram of the basic structure is given in Figure 8.1.1a. (West 1993). This shows the location of veins and arteries in the dermis, which in turn is separated from the skin surface by the stratum corneum. Thus, even for *in vitro* monitoring (e.g., identification of different cells in pathological samples, etc.), the conditions to be monitored are complex. For *in vivo* monitoring, there are additional complexities such as regular pulsations by which blood flows through the veins and arteries, carrying with them a range of hemoglobin derivatives (West 1993). Optical monitoring of such a complex structure with its different functionalities is therefore technologically demanding.

The complex tissue structure interacts with light in a number of ways, including the processes of optical reflectance, absorption, and scattering (Chapter 7), and can cause optical interference effects because the various layers may act as interference cavities (Chapter 6, Sections 6.2.1 and 6.2.3). Some chemical components (e.g., glucose) may be optically active and may affect the state of polarization of light (Chapter 6, Section 6.3.4, Polychromatic Rotary Activity).

The reflectance, absorption, and scattering effects vary substantially with different tissue and blood components (Figure 8.1.1b) and also with the wavelength of the light (Van Gemert et al. 1989). Within the wavelength range 200–900 nm, the stratum corneum, epidermis, and dermis have relatively high scattering coefficients and lower absorption coefficients—particularly at longer wavelengths. Consequently, the spectrum of white light passing through such tissue can be affected by the various components and become biased towards the longer wavelengths. This means that the light emerging from such tissue carries valuable information about both the tissue and the blood in the veins and arteries, which could be extracted with appropriate techniques. The chromatic approach provides such possibilities and also deals with artifacts such as skin pigmentation, which could interfere with diagnosis.

Applications of chromatic techniques for monitoring biological tissue and blood via reflectance, absorption, scattering, and interference of polychromatic light are described in the sections that follow. An example of addressing the reflectance and scattering of polychromatic light from skin tissue (Jones et al. 2000) is the monitoring of bilirubin levels (Jacques et al. 1997) in neonates to reduce the need for excessive sampling of blood from the limited

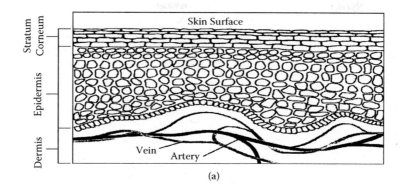

(a)

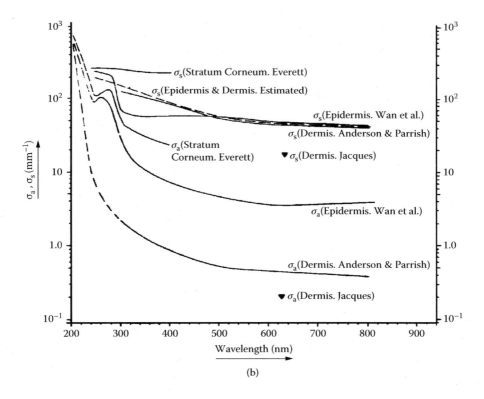

(b)

FIGURE 8.1.1
Skin tissue properties: (a) basic structure, (b) experimental absorption (σa) and scattering (σs) coefficients versus wavelength. ((a) From West, I. P. (1993). Pulse Oximetry for Magnetic Resonance Scanner Environments. Ph.D. thesis. University of Liverpool; (b) from Van Gemert, M. J. C., Jacques, S. L., Sterenborg, H. J. C. M., and Star, W. M. (1989). Skin optics. *IEEE Trans. Biomed Eng* 36: 1146–1154. With permission.)

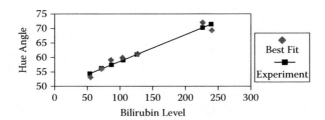

FIGURE 8.1.2
Chromatic H as a function of bilirubin level for a number of neonates. (From Jones, G. R., Russell, P. C., Vourdas, A., Cosgrave, J., Stergioulas, L., and Haber, R. (2000). The Gabor transform basis of chromatic monitoring. *Meas. Sci. Tech.* (11), No. 5, 489–498. With permission.)

amount available in them. In this case, through the choice of appropriate chromatic processors (Chapter 2, Section 2.2), a linear relationship between the chromatic H parameter and bilirubin level is achievable (Figure 8.1.2).

Addressed chromatically, the pulsatile component of absorbed and scattered polychromatic light can be used for pulse oximetry (Section 8.2), whereas reflected and scattered light can provide indications of tissue oxygenation (Section 8.3). Both applications can be addressed with optical fiber transmission. Chromatic monitoring of reflected and scattered polychromatic light can be used for *in vivo* detection of skin cancer (Section 8.4). Also, chromatic processing of two-dimensional images formed by reflected and scattered light can be used for the *in vitro* detection and identification of different types of white blood cells (Section 8.5).

Chromatic techniques can also be used to improve the resolution of low-coherence optical coherence tomographic (OCT) Fercher et al. 2003) images of microscopic tissue layers (Section 8.6).

8.2 Chromatic Pulse Oximetry

8.2.1 Introduction

Pulse oximetry is concerned with measuring blood oxygen levels *in vivo* with portable instrumentation that monitors the absorbance and scattering of light by various forms of hemoglobin (Severinghaus and Kelleher 1992; Wukitsch et al. 1988; Griffiths et al. 1988). The probing light may be delivered to and received from a well-defined area of tissue, for example, on a finger, using optical fibers, which also provide immunity to electrical interference. The oxygen level is quoted as a percentage of the hemoglobin that is in the form of oxy-hemoglobin (HbO) rather than reduced hemoglobin (RHb), according to the following equation (West 1993):

$$SpO_2 = cHbO/(cHbO + cRHb) \qquad (8.2.1)$$

The variations of the extinction coefficients of oxy-hemoglobin and reduced hemoglobin with wavelengths are different (Figure 8.2.1a) but have a cross-over point at approximately 805 nm (isobestic point) (Van Assendelft 1970).

Conventionally, oxygen saturation (Equation 8.2.1) is determined from the ratio of the intensities of optical signals at two wavelengths, one on either

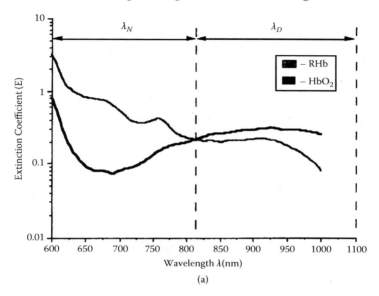

(a)

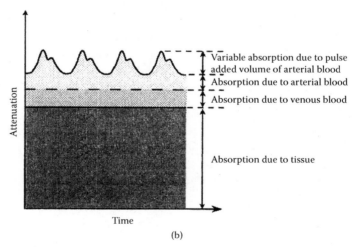

(b)

FIGURE 8.2.1
Pulse oximeter operation. (a) HbO_2 and RHb extinction curves defining lambda (A), lambda (D); (b) light absorption in an arterial tissue bed. ((a) From Van Assendelft, O. W. (1970). Spectrophotometry of Hemoglobin Derivatives. *Royal Van Gorcum*, Assen, Netherlands; and Charles C Tomas, Springfield, Ill. With permission; (b) from West, I. P. (1993). Pulse Oximetry for Magnetic Resonance Scanner Environments. Ph.D. thesis. University of Liverpool.)

side of the isobestic point. This ratio is calibrated against the oxygen level measured *in vitro* by other means.

In vivo, there is blood flowing through veins and arteries (Figure 8.1.1) in a pulsatile manner, so the blood-related signals are time varying, whereas the surrounding tissue-related signals are steady (Figure 8.2.1b). Variations in optical intensity due to artifacts such as optical fiber movement, source intensity changes, etc., may be overcome by using the ratio of the pulsatile to the steady signal (West 1993).

The distributed nature of the wavelength dependence of the extinction coefficients of HbO and RHb (Figure 8.2.1a) means that chromatic, as opposed to narrow wavelength, techniques can offer advantages for monitoring. For example, improved signal strengths without overloading the tissue with excessive optical power at restricted wavelengths is possible (West 1993).

8.2.2   Distimulus Chromatic System

The two-part nature (λ_N, λ_D, Figure 8.2.1a) of the RHb, HbO spectral range lends itself to using only two nonorthogonal processors (Distimulus System, Chapter 3, Section 3.3.2) to simplify optical fiber instrumentation. If the pulsatile outputs from each processor are $i(\lambda_N)$, $i(\lambda_D)$, and their steady components are $I(\lambda_N)$, $I(\lambda_D)$, then the dominant wavelength (λ_d) of the captured light, normalized to accommodate artifacts (fiber movement, etc.), may be written (West 1993) as

$$\lambda_d = R = [i(\lambda_N)/i(\lambda_D)][I(\lambda_D)/I(\lambda_N)] \qquad (8.2.2)$$

λ_d may be calibrated against blood oxygen saturation. For a system based on a white LED source in conjunction with two particular optoelectronic detectors (West 1993), produces the characteristic shown in Figure 8.2.2a.

Chromatic signals received from such a distimulus optical fiber unit for a subject undergoing a routine magnetic resonance scan (for which the use of optical fibers has radio frequency immunity advantages) are shown in Figure 8.2.2b (West 1993).

The pulsatile (n,d) and steady (N,D) outputs from the distimulus unit are shown along with the dominant wavelength (R, Equation 8.2.1) and the oxygen saturation level (SpO_2) determined from the calibration graph. The subject's pulse rate (P), which produces the pulsatile aspect of the oximeter signal, is also shown.

8.2.3   Skin Pigmentation Effects

Investigations of the possible effects of skin pigmentation on pulse oximetry accuracy have been inconclusive (West 1993); some reports show erroneously high readings (3–5%) for Negroid subjects, whereas others have found no significant effects (Emery 1987, Gabrilczyk and Buist 1988). Optical transmission spectra through the fingers of subjects from different ethnic groups (West 1993) reveal variations in the intensities of light transmitted but only

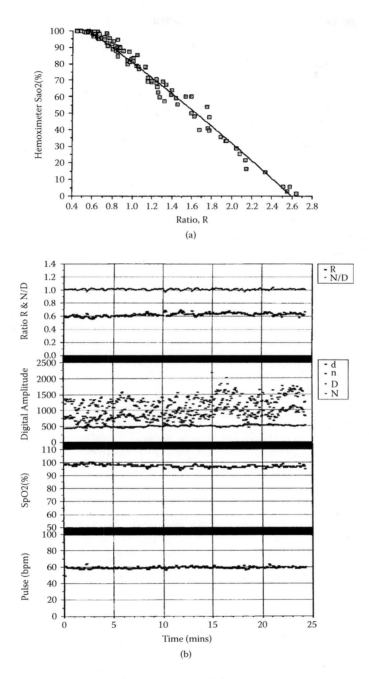

FIGURE 8.2.2
Distimulus pulse oximeter test results: (a) oxygen concentration versus dominant wavelength, R, (b) data logged for a subject during a magnetic resonance scan. (From West, I. P. (1993). Pulse Oximetry for Magnetic Resonance Scanner Environments. Ph.D. thesis. University of Liverpool.)

small differences in wavelength distributions apart from those from Negroid skin types (Figure 8.2.3a).

The theoretical dominant wavelength variation with oxygen levels determined with a white LED chromatic distimulus system shows little difference between those for Caucasians and Negroid subjects (Figure 8.2.3b), these being the ones with the greatest spectral differences (Figure 8.2.3a).

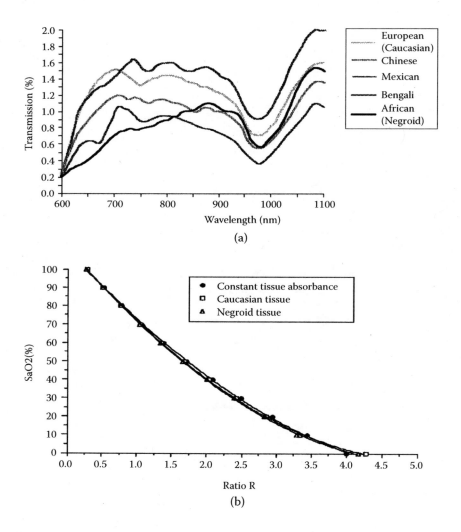

(a)

(b)

FIGURE 8.2.3
Effect of skin pigment: (a) transmission spectra through the fingers of subjects of different ethnic origins, (b) LED system theoretical calibration curves for ● constant tissue absorbance (over LED bandwidths), □ Caucasian tissue absorbance, Δ Negroid tissue absorbance. (From West, I. P. (1993). Pulse Oximetry for Magnetic Resonance Scanner Environments. Ph.D. thesis. University of Liverpool. With permission.)

The assumption of a wavelength-independent tissue absorbance only has a small effect, mainly below 80% oxygen saturation.

The immunity shown by these results to different tissue pigments is valid for particular combinations of optical sources and detectors of the chromatic system. Other system components (e.g., tungsten halogen source) having spectra extending further into the infrared can show marginally different responses for Caucasian and Negroid tissue, but only for oxygen saturation levels less than 65% (West 1993).

8.2.4 Tristimulus Chromaticity and Dyshemoglobins

Dyshemoglobins are hemoglobin species that render the hemoglobin molecule either permanently or temporarily unable to bind to oxygen (West 1993; Ralston et al. 1991). The most common examples of dyshemoglobins are carboxy-hemoglobin (COHb) and Met-hemoglobin (MetHb). Although both species are normally less than 2% in concentration, COHb can be elevated in the case of individuals who smoke or subjects exposed to fires or vehicle exhaust fumes (Freeman and Perks 1990; Ralston et al. 1991). Readings from conventional pulse oximeters can be misleading in the presence of such dyshemoglobins because they ignore the spectral effects of the latter (Ralston et al. 1991). The clinically significant parameter in such cases is the "fractional saturation" $FSaO_2$ (West 1993):

$$FSaO_2 = cHbO/(cHbO + cRHb + cCOHb + cMetHb) \qquad (8.2.3)$$

The spectral complexity produced by the presence of COHb and MetHb in addition to HbO and RHb is substantial and difficult to resolve with systems based on only two detectors (Figure 8.2.4a) (Van Assendelft 1970). However, a tristimulus chromatic system (three detectors, see Chapter 3, Section 3.3.2 (d)) has in principle the capability of resolving the problem. The outputs of the three processors of a tristimulus chromatic system, both pulsatile and steady, may be transformed into two ratiometric parameters, each in the form of the distimulus relationships 3.20. 3.21 (West 1993).

$$R1 = [i (X)/i (Z)] [I (Z)/I (X)] \qquad (8.2.4)$$

$$R2 = [i (Y)/i (Z)] [I (Z)/I (Y)] \qquad (8.2.5)$$

where i (X), i (Y), i (Z) ($[r_0, g_0, b_0,$ of Equations 3.20, 3.21]) are the pulsatile outputs from the processors X, Y, Z (R, G, B of Equations 3.20, 3.21) and I(X), I(Y), I(Z) ($[R_0, G_0, B_0,$ of equations 3.20, 3.21]), their steady outputs. Synonymous with Equation 8.2.1, these relationships may be regarded as representing dominant wavelengths from two different chromatic dimensions relating to the spectral regions covered by X, Z and Y, Z.

When R1 and R2 are each plotted against the oxygen saturation for various proportions of HbO, RHb, and COHb, graphs of the form shown in Figure 8.2.4b (West 1993) are produced. These show the extent to which the different concentrations of COHb affect the HbO calibration curve

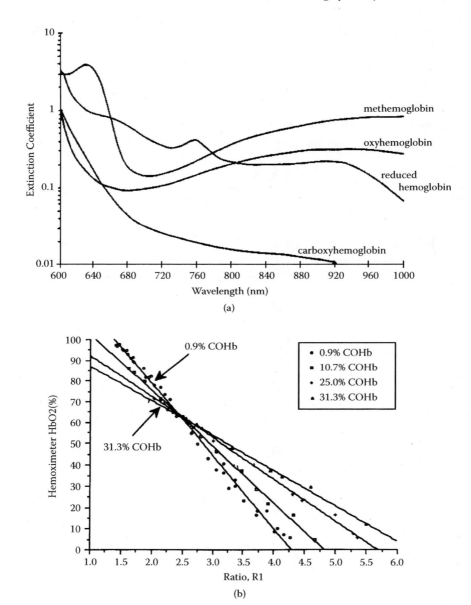

(a)

(b)

FIGURE 8.2.4
Addressing dyshemoglobins with tristimulus chromaticity: (a) extinction coefficient versus wavelength for HbO, RHb, COHb, MetHb, (b) oxygen saturation versus R1 for various COHb concentrations (c) R1:R2 Chromatic Diagram for Various HbO and COHb concentrations. ((a) From Wukitsch, M. W., Petterson, M. T., Tobler, D. R., and Pologe, J. A. (1988). Pulse oximetry: Analysis of theory, technology, and practice, *J. Clin. Monit. Comput.*, 4, 4: 290–301. Permission needed; (b) from West, I. P. (1993). Pulse Oximetry for Magnetic Resonance Scanner Environments. Ph.D. thesis. University of Liverpool.)

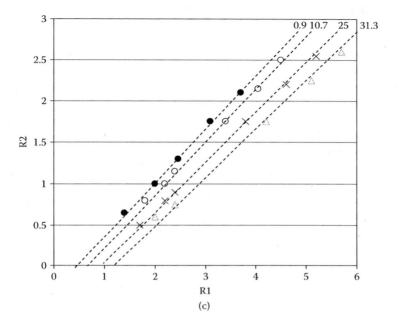

FIGURE 8.2.4
(Continued).

(the region in which the curves overlap being less sensitive to COHb, different for R1 and R2, and lower than regions of most clinical interest). The calibration curves may be represented by equations of the following form (West 1993):

$$FSaO_2 = mn.Rn + cn \qquad (8.2.6)$$

where mn and cn are functions of the COHb concentration and $n = 1,2$. The inverse problem of extracting HbO and COHb concentrations from measured values of R1, R2 is difficult but can be made more tractable by recourse to a chromatic diagram of R1 versus R2 (Figure 8.2.4c). When transformed into this domain, the data lie on a series of parallel straight lines, each corresponding to a specific COHb concentration and obeying the equation

$$R2 = mr.R1 + cr \qquad (8.2.7)$$

where mr is independent of COHb, and cr is a function of COHb. Consequently, because mr is known *a priori* from calibration, cr can be determined for each pair of measured values of R1 and R2, and the COHb concentration extracted. Once the COHb concentration is known, appropriate values of m1 and c1 corresponding to the COHb concentration (previously determined by calibration, Figure 8.2.4b, Table 8.2.1) can be inserted into Equation 8.2.6 to yield the HbO concentration:

$$(HbO = m1.R1 + c1)$$

TABLE 8.2.1

Coefficients m1 and c1 for
Various COHb Concentrations

COHb (%)	−m1	c1
0.9	34.43	147.67
10.7	26.77	129.27
25	19.66	111.84
31.3	16.58	103.62

Source: West, I. P. (1993). Pulse Oximetry for Magnetic Resonance Scanner
Environments. Ph.D. thesis. University of Liverpool.

8.3   Tissue Blood Oxygenation (TBO)

8.3.1   Introduction

Tissue oxygenation is defined (Glomon 2003) in terms of a parameter α given
by

$$\alpha = \frac{X - Oc}{X_N - Oc} \tag{8.3.1}$$

where X, X_N, and Oc are the spectral transmission factors at a wavelength of
560 nm of the tissue under observation, under normal oxygenation conditions, and at total occlusion, respectively. For total and normal occlusion, $\alpha = 0$
and 1, respectively; $\alpha > 1$ indicates an overshoot of TBO.

One approach for acquiring such information is to illuminate the tissue
with polychromatic light via optical fibers and to monitor changes in the optical spectrum of the returned light (Nicholson 2002). The spectral changes
(Figure 8.3.1) are distributed throughout the wavelength range. The conventional approach for extracting information has been to identify a wavelength
at which the signal amplitude change caused by the system condition is maximum and measure that localized change. Such an approach is susceptible to
poor signal-to-noise (S/N) properties and is not capable of indicating if there
has been intervention from other effects. However, the distributed nature of the
spectral changes represents a condition addressable by a chromatic approach.
This can alleviate such difficulties by improving the S/N ratio because of the
integrated nature of the processing and can be indicative of intervention from
other effects by affecting the cross correlation between H, S, and L.

8.3.2   Basic Chromatic Analysis

Passage of the spectral data from Figure 8.3.1 through three chromatic processors yields normalized R, G, B values as a function of $\propto$.

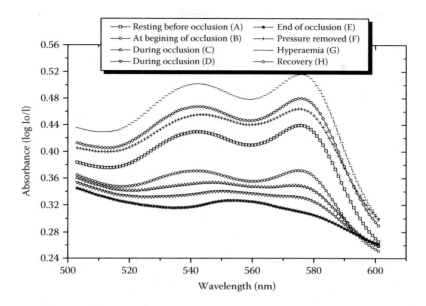

FIGURE 8.3.1
Optical spectra from a thumb for different degrees of blood occlusion. (From Nicholson 2002, Private communication. With permission.)

Values of the chromatic parameters *H,L,S* are derived from the *R,G,B* values, respectively, by using Equations 1.2 to 1.7, Chapter 1, except that the S parameter definition is modified to

$$S' = [(R + B) - G]/[R + B = G] \tag{8.3.2}$$

The results showing the variations of H,L,S with oxygen-level parameter (α) are given in Figure 8.3.2a–c. These show L to have a fairly linear variation throughout the range $0 < (\alpha) < 1.8$, S in the range $0.4 < (\alpha) < 1.3$. Because L is susceptible to various optical intensity–affecting aberrations, it is advantageous to use values normalized with respect to the normal occlusion level $(\alpha) = 1$.

The results in Figure 8.3.2 are for separate tests on two different thumbs and show good agreement to confirm the consistency of the results.

8.3.3 Chromatic Vector Representation

Each chromatic parameter (H,L,S) shows different sensitivities to and relative magnitudes with α in various ranges. For example, L increases with α and so has relatively low values at low α, whereas H and S decrease with α and have high values at low α. H values are generally more reliable at higher S, and L is more susceptible to aberrations at low values. There is therefore merit in combining the chromatic parameters vectorially according to Equation 1.13,

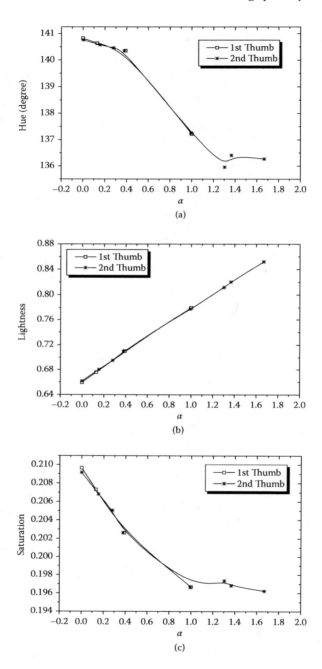

FIGURE 8.3.2
Hue, saturation, and lightness as functions of (α) for thumb tissue spectra. (a) hue, (b) lightness, (c) saturation (results for two different thumbs). (From Glomon, L. (2003). Source-Based Chromatic Methodology for optical fiber sensor systems. Ph.D. thesis, University of Liverpool.)

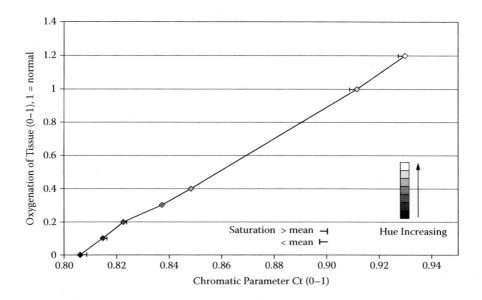

FIGURE 8.3.3
Variation of chromatic parameter CT (Equation 8.3.3) with oxygen occlusion. (From Rallis, I., Glomon, L., Deakin, A.G., Spencer, J.W., and Jones, G.R., Polychromatic Monitoring of Complex Biological Systems, Proc. Complex Systems Monitoring Session of the International Complexity, Science and Society Conference [Liverpool 2005]. With Permission.)

Chapter 1, to yield

$$C_T = \sqrt{a_H \cdot H_N^2 + a_L (b_L - L_N)^2 + a_S (b_S - S_N)^2} \qquad (8.3.3)$$

where a_H, a_L, a_S are scaling constants and H_N, L_N, S_N are normalized values of H,L,S, respectively. Because H and S decrease (dH/dS positive) and L increases (dH/dL negative), bS = 0 and bL = 1 (see Equations 1.14–1.16, Chapter 1) (Rallis et al. 2005). The relationship between C_T and α is monotonic as shown in Figure 8.3.3, enabling the oxygen level to be unambiguously determined throughout the measurement range.

8.4 Skin Cancer Diagnosis

8.4.1 Introduction

Many technical advances and new methodologies for early detection of cancer have been developed in recent times, but there are still considerable challenges in the frames of precise diagnosis and differentiation of cutaneous malignant lesions. Skin cancer is one of the most widespread tumors worldwide. The most malicious form of cutaneous neoplasia is malignant melanoma (MM), which has poor patient prognosis. As such, it continues to be an important problem of social health worldwide.

During the last few years, the frequency of potentially lethal melanocytic neoplasia has increased worldwide (Mackie et al. 1997; Greenlee et al. 2000; Wingo et al. 1999), and so has the mortality rate (Greenlee et al. 2000). Patients' survival with MM lesions is directly related to early diagnosis of this disease. Tumors detected in the early stage have good prognosis and low risk for metastasis development (NIH 1992, Sahin et al. 1997).

Most dermatologists rely on their practical experience in visual evaluation of pigmented lesions. It is ascertained that diagnostic accuracy depends on the clinical experience of the specialist (Morton and Mackie 1998). In general practice, where such experience is low, diagnostic accuracy is quite bad (Bedlow 1995). In the early stages of the disease, diagnosis is difficult even for experienced clinicians.

Of many techniques available for addressing the detection of skin lesions, reflectance spectroscopy has been used for differentiating between the pigmentation of various kinds of lesions (Wallace et al. 2000a, 2000b; Marchesini et al. 1991, 1992; Farina et al. 2000). To benefit fully from reflectance spectroscopy's advantages, one needs to relate the spectral features with the morphology and biochemical composition of the tissue investigated. Various other skin tissue characteristics, such as the melanin content in the epidermal layer, hemoglobin derivatives in the dermis, and their influence on the reflectance spectra need to be taken into account (Sections 8.1), thus adding to the complexity of the conditions to be monitored. Chromatic techniques have the potential for providing the means for quantifying the tissue spectral characteristics selectively under such conditions and for producing an assistive automated diagnosis.

8.4.2   Basis of Diagnosis Method

Figure 8.4.1a shows the reflectance spectra from healthy tissue for each of two patients (P1h, P2h), along with the reflectance spectra from lesions for each patient (P1s, P2s). In both cases, the "lesion" spectrum is of a lower intensity and appears darker and more evenly distributed over the wavelength range than the "healthy" spectrum. These differences in spectral features and intensity distributions are typical of healthy and diseased tissues, although there may be variations superimposed for various particular individuals.

There are also detailed differences between the spectral reflectance of various types of diseased tissues (Borisova et al. 2005) and variations for each type of disease for different individuals. Consequently, if such aberrations could be accommodated, the reflectance spectra could provide a means for distinguishing between healthy and diseased tissues and also between different diseases if a method for quantifying the relevant differences between the spectra could be established. The difference between two spectra may be emphasized via the ratio of one spectrum to the other and the diagnosis

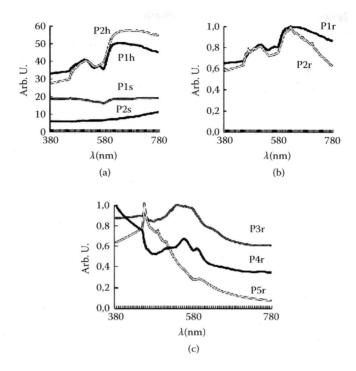

FIGURE 8.4.1
Examples of spectra from the tissue of various patients. (a) reflectance spectra from healthy skin (Ph) and from lesions (Ps) of two patients (P1, P2) with the same disease, (b) spectra calculated as a ratio between "healthy" and "diseased" spectra (Pr) for patients P1, P2, (c) ratio spectra (Pr = healthy/diseased) for three patients (P3, P4, P5) with different diseases.

may then be made by applying chromatic processing to this ratio signal (Pavlova et al. 2004). A decision chart for distinguishing diseased tissue from healthy tissue and for distinguishing between different diseases is given in Figure 8.4.2.

8.4.3 Calibration Procedures

The procedures for providing classification of various tissue conditions are as follows.

The more reliable spectral differences between various tissue samples are in the distributed nature of the spectra rather than amplitude, although the latter is often a coarse indicator. It is therefore advantageous to amplify the differences between two spectra being compared by determining the ratio of one with respect to the other, i.e., by evaluating ($r(\lambda) = q(\lambda)/h(\lambda)$) at each wavelength, λ, where $q(\lambda)$ and $h(\lambda)$ are the absolute amplitudes at each λ of the suspect and healthy tissues, respectively. For example, applying this

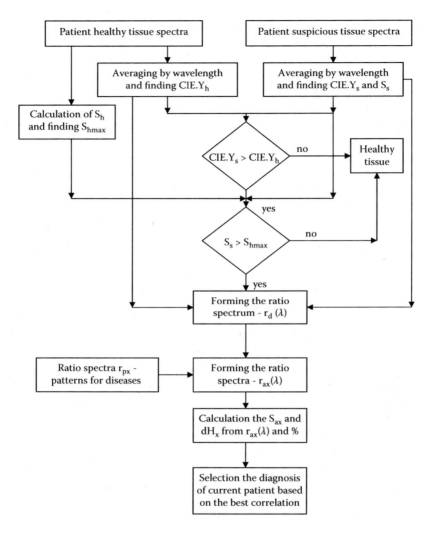

FIGURE 8.4.2
Decision diagram for identifying diseased tissue.

procedure to the spectra shown in Figure 8.4.1a leads to the ratio spectra shown in Figure 8.4.1b.

Each of the ratio spectra $r_d(\lambda)$ of Figure 8.4.1b may then be addressed by three wavelength-domain chromatic processors from whose outputs the chromatic parameters H and S can be evaluated (Figure 8.4.2). Because such absolute spectra have similar wavelength distributions, the ratio spectra are quite evenly distributed with respect to wavelength so that, in general, $S_e \sim 0$ and H_e are fixed values.

The same type of wavelength-domain chromatic processors may be applied to a range of samples of healthy tissue to establish average and maximum values of S_{hmax} corresponding to the healthy tissue.

The ratio spectrum ($r_{px} = d(\lambda)/h(\lambda)$) of each diseased tissue may be formed from the absolute spectra of the diseased tissue ($d(\lambda)$)and the healthy tissue ($h(\lambda)$). Examples of such ratio spectra (patterns) for six different types of diseased tissues are shown in Figure 8.4.3a–c. The diseased tissues are base cell carcinoma, compound nevus, dermal nevus, dysplastic nevus, malignant melanoma1, and malignant melanoma 2 (Pavlova et al. 2004).

New spectra r_{ax} calculated as a ratio between ratio spectra r_{px} and r_d ($r_{ax}(\lambda) = r_{px}(\lambda)/r_d(\lambda)$) possess the same peculiarities as the other ratio spectra—the S_a and H_a for quite even distributions with respect to wavelength are $S_a \sim S_e$ and $H_a = H_e$. Thus, the small value of S_a and the short distance dH between H_e and H_a show a higher degree of similarity than the higher ones. The minima from among the S_{ax} and dH_x correspond to particular diseases. For such

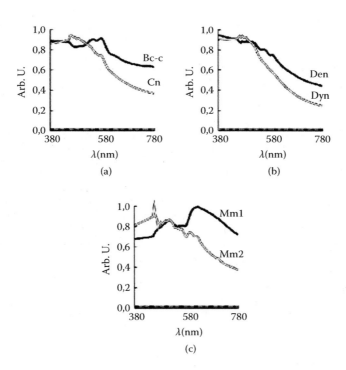

FIGURE 8.4.3
Ratio-spectra for various types of diseases. (a) Bc-c base-cell carcinoma; Cn compound nevus, (b) den dermal nevus; Dyn dysplastic nevus, (c) Mm1 malignant melanoma 1; Mm2 malignant melanoma 2. (From Pavlova, P., Borisova, E., and Avramov. L. (2004). Automation of the Skin Cancer Diagnostics on the Basis of Reflectance Spectroscopy. *Proc. 9th Natl. Conf. Biomed. Phys. Eng.* 260–265.)

purposes, it is useful to assign a percentage value to the suspect tissue S_{ax} and dH_x parameters as a complement to 100%.

8.4.4   Test Procedure

The manner in which the methodology is deployed for identifying diseased tissue and the type of disease is as follows.

The ratio spectra of suspect tissue samples are formed (Figure 8.4.2). Examples of such ratio spectra for five different patients (P1–P5) are given in Figure 8.4.1b and c.

An estimate is made as to whether the suspect tissue is diseased according to a cross correlation between two criteria:

- The CIE Y parameter determined from the outputs of the three chromatic processors (Equation 3.5, Chapter 3) is evaluated for the suspect tissue (CIE Y_s) and compared with that of healthy tissue (CIE Y_h). A lower value of Y for the suspect tissue (indicating lower signal intensity at longer wavelengths—compare Figure 8.4.1a) suggests that the tissue could be diseased.
- A value of S_s (Equation 1.4, Chapter 1) for the suspect tissue, which is higher than that of the maximum healthy tissue for the individual ((S_{hmax}), increases the possibility of the tissue being diseased (Figure 8.4.2).

Once the possibility of the suspect tissue being diseased according to the foregoing tests is established, the ratio spectrum r_d for the individual is compared with those of each type of diseased r_{px} (Figure 8.4.2) as new ratio spectra r_{ax}. The highest percentage score for S_{ax} and dH_x indicates the most likely type of tissue disease.

To mitigate the uncertainty caused by S_{ax} and dH_x giving contrary indications, a third parameter $(S + H)$ is considered. In the event of all three parameters giving contrary indications, the S_{ax} value dominates.

8.4.5   Examples of Deployment

The deployment of the methodology is illustrated by its application in diagnosing tissues from five different patients, P1–P5, whose ratio spectra have been presented in Figures 8.4.1b and c.

The S, H, and $(S + H)$ values determined as described in Section 8.4.4 are summarized in Table 8.4.1 for the five patients (P1–P5) with respect to tests for each of the six abnormal tissue conditions for which ratio spectra were determined (Figure 8.4.3a–c). For patient P1, the highest percentage values for S, H, and $(S + H)$ (93, 79, and 85, respectively) correlate to indicate the most

TABLE 8.4.1

Percentage Affiliation of Patients to Each Disease Pattern

Patterns	P1			P2			P3			P4			P5		
	S	H	Σ	S	H	Σ	S	H	Σ	S	H	Σ	S	H	Σ
							Percentage								
Base-cell carc.	65	77	71	67	76	71	95	97	96	78	72	75	29	74	51
Comp. nevus	41	78	59	44	77	60	64	79	71	84	73	78	42	73	57
Dermal nevus	50	78	64	52	78	65	77	82	79	98	61	79	36	74	55
Dysp. nevus	32	78	55	34	78	56	50	79	64	66	77	71	51	73	62
Malign. mel. 1	93	77	85	96	73	84	70	78	74	54	75	64	20	75	47
Malign. mel. 2	52	79	65	54	78	66	80	84	82	94	77	85	35	74	54
Real diagnosis	Malignant melanoma			Malignant melanoma			Base-cell carcinoma			Dermal nevus			Dysplastic nevus		

Note: Pi—patient; S—saturation distance; H—hue distance; Σ—average total S and H—distance.

Source: Pavlova, P., Borisova, E., and Avramov. L. (2004). Automation of the Skin Cancer Diagnostics on the Bases of Reflectance Spectroscopy. *Proc. 9th Natl. Conf. Biomed. Phys. Eng.* 260–265.

probable tissue disease to be malignant melanoma 1. For patient P3, the highest percentage values for S, H, and (S + H) (95, 97, and 96, respectively) suggest base cell carcinoma. The distinction between dermal nevus and dysplastic nevus diseases (patient P4) is not clear because of the similarity between their ratio spectra (Figure 8.4.3b).

8.4.6 Conclusions

The utilization of reflectance ratio spectra provides a useful means for enhancing chromatic information extraction.

As stated, melanoma incidence and mortality rates are on the increase in many countries. There is much evidence in clinical practice to show that the standard biopsy could be one reason for dissemination of cancer cells and that it is not advisable for it to be applied. In this context, the development of noninvasive, rapid, and reliable methods attains significant importance.

The deployment of chromatic methods in conjunction with the ratio spectra of various tissue conditions shows potential for identifying diseased tissues and for distinguishing between certain groups of tissue diseases.

The method described allows creation of an entirely automated diagnostics, overcoming subjectivity, and could be a valuable addition to the present cancer detection techniques. However, discrimination between some diseases that have similar but not identical ratio spectra requires further chromaticity-based developments.

8.5 Differential Blood Counting

8.5.1 Introduction

Differential blood counting is a method for quantifying the different nature of various types of white blood cells in a peripheral blood sample. These are lymphocytes, monocytes, basophils, and neutrophils (Dobreva and Meshkov 1984). There are two techniques currently employed: the first is based on laser flow cytometry with impedance counting, and the second is microscope-based. The latter involves analyzing images derived from smears, but the former does not involve imaging. The images obtained using the microscope can be saved and compared for the purpose of studying the pathology or phase of cell evolution, tracing the history of diseases, or automating the diagnostics if a computer is used for processing.

In the image case, the visual or automatic identification is based on the coloring of the cellular elements after staining the smears (Dobreva and Meshkov 1984). However, these colors are variable and change because of differences in the chemical staining process or in the cell, or because of morphometric features, illness, or a person's blood specifics.

8.5.2 Analytical Basis

The aforesaid complications make automatic analysis more difficult. However, the variability problem can be overcome if only the chromatic parameters (hue, saturation, value, Chapter 3, Section 3) of the color, which are significant for the required identification, are employed. Experience suggests that the order of significance of the chromatic parameters for distinguishing the required features is H, S, V (Pavlova and Staneva 2005). The image of a single cell may be extracted and rules for distinguishing a cell type established using only specific features of the histograms based on only the chromatic parameters of S and H (Pavlova, et al. 1996; Pavlova 2001). The specific features include the size of the single cell image, the ratio between the number of pixels forming the cytoplasm and nucleus, and the ratio between the number of pixels forming cytoplasm and granules.

8.5.3 Chromatic Processing Method

The colors of the cellular elements are produced by the May–Grunwald–Romanowski–Giemsa standard cytochemical staining. The color differences and similarities produced by this protocol are shown in Table 8.5.1 for the

TABLE 8.5.1

Color of the Cellular Elements in a Smear Stained by May–Grunwald–Romanowski–Giemsa

Cell Elements	Cytoplasm					Nucleus and partic.			Granules			
Color Cell Type	Red	Pink-Brown	Pink	Gray-Blue	Dark Blue	Red-Violet	Dark Violet	Pink	Brown	Red	Onto the Nucleus	Outside the Nucleus
Rubriblasts	No	Yes	No	No	No	Yes	No	No	No	No	No	No
Rubricytes	No	Yes	No	No	No	Yes	No	No	No	No	No	No
Erytrocytes	No	Yes	No	No	No	No	No	No	No	No	No	No
Basophils	No	No	Yes	No	No	Yes	Yes	No	No	No	Yes	No
Eosinophils	No	No	Yes	No	No	Yes	No	No	Yes	No	No	Yes
Neutrophils	No	No	Yes	Yes	No	Yes	No	Yes	No	No	No	Yes
Monocytes	No	No	No	Yes	No	Yes	No	No	No	No	No	No
Limphocytes	No	No	No	No	Yes	Yes	No	No	No	Yes	No	Yes
Thrombocytes	Yes	No	No	No	No	No	No	No	No	No	No	No

Source: Pavlova, P. (2001). Creation of a pattern features model for identification in computer differential blood count. *Comp. Rend. de l'Acad. Bul. Sci* 54: 29. With permission.

cellular elements of interest as well as all other possible cells occurring in a single smear (Dobreva and Meshkov 1984).

Images of white blood cells derived from peripheral blood smears need to be captured via a microscope by a camera connected to a computer. The image is transformed into H, S, V chromatic spaces with software. Using standard operations (Rogers 1985), H and S pixel histograms are produced covering the H range 0–359 degrees in 1° steps and the S range 0–1.0 in 0.01 steps. The cell and relevant images are separated from the rest of the image with software according to the subranges of the H and S histograms within which they reside and using the information contained in Table 8.5.1.

This process yields the following characteristics for distinguishing a single cell from the background:

1. The cytoplasm and the background are almost equally saturated (same S values), and their pixels form the maximum on the S histogram. However, the H values of the cytoplasm and the background locate them in different subranges of the H range.

2. The elements most deeply affected by cytochemical staining have a maximum saturation (S), and the corresponding pixels have a second local maximum on the S histogram.

These features are incorporated into an algorithm for automatically identifying and locating a single cell into a field of view, compensating for differences of color, and ensuring that all elements that are significant for identifying particular types of cells are confined within the H range $220 < H < 350$ degrees.

8.5.4   Chromatic Signatures

A graphic image of a single cell, including cellular elements identified with the procedure described in Section 8.5.3, is shown in Figure 8.5.1a along with the S and H histograms of the image. The S histogram shows two peaks, one pronounced and the other less so. The H histogram has a skewed profile with a tail extending to higher H values.

Figures 8.5.1b and c correspond, respectively, to the S, H histograms of the inner and outer structures of the cell. These show that the outer structure produces the smaller, lower, S-value peak on the S histogram, and the higher H components on the H histogram.

Single-cell histograms of the form shown in Figure 8.5.1 enable average H histograms to be established for basic leukocyte cell types. Figure 8.5.2 shows such averaged histograms after filtering for a range of cell types—lymphocytes, monocytes, basophils, eosinophils, and neutrophils. Each cell type is distinguishable from the different profiles of the histograms, providing

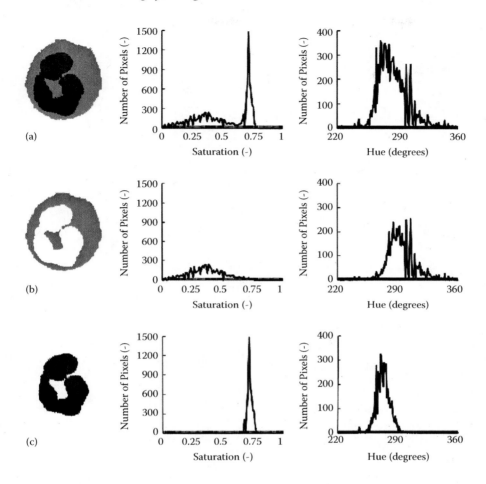

FIGURE 8.5.1

Image and histograms of S, H—pixel numbers for a separated cell. (a) overall histogram, (b) histogram of nonsaturated part of the cell, (c) histogram of the saturated part of the cell nucleus. (From Pavlova, P. (2001). Creation of a pattern features model for identification in computer differential blood count. *Comp. Rend. de l'Acad. Bul. Sci* 54: 29–34. With permission.)

thereby a means for identifying different cells from the value of a single chromatic parameter H.

In principle, it is possible to further quantify the H histograms of different cells (Figure 8.5.2) by applying secondary chromatic processors, as described in Chapter 1, Section 1.3.2 (d), to each of the histograms so that the cell type can be defined in terms of the values of the three secondary chromatic processors.

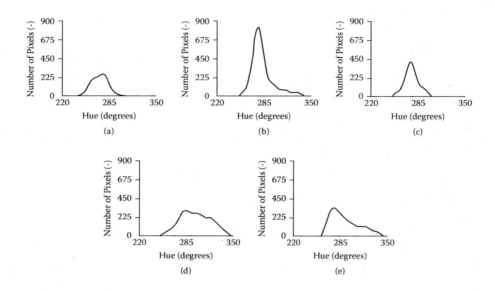

FIGURE 8.5.2
Averaged filtered histograms of the different leukocyte types. (a) lymphocytes, (b) monocytes, (c) basophils, (d) eosinophils, (e) neutrophils. (From Pavlova, P. (2001). Creation of a pattern features model for identification in computer differential blood count. *Comp. Rend. de l'Acad. Bul. Sci* 54: 29–34. With permission.)

8.6   Optical Coherence Tomography (OCT)

8.6.1   Introduction

Optical coherence tomography (OCT) (Fercher et al. 2003) is a method being researched for monitoring subsurface structures of biological tissue for medical applications such as ophthalmology and endoscopy. Such subsurface structures are not visible through the surface with conventional optical imaging and are complex in nature. The typical structure of human tissue has been shown in Figure 8.1.1 (Section 8.1), which illustrates the complexity of the structure (epithelium, lamina propria, muscularis mucosa, and submucosa) and highlights some of the more significant tissue interfaces.

OCT produces a cross-sectional image from the interference between light waves reflected from the various subsurface layers and light reflected from a moving reference (Chapter 6, Section 6.2.3). The type of phase-domain interference signals produced have been shown theoretically in Figures 6.2.3.1 and 6.2.3.2 (Chapter 6). The manner in which chromatic processing can be applied to such phase-domain interference signals has also been described in Chapter 6, Section 6.2.3.

In the case of OCT signals, the interference pattern may be regarded as the addition of a series of interferograms derived from reflections from each tissue interface. This addition is simple if the subsurface interfaces are sufficiently spaced apart, but becomes complex should the separation fall below the coherence length of the light source, which results in the component interferograms overlapping.

The typical light sources used for OCT have a low temporal coherence (Chapter 6), being typically of the order of 10 μm, similar to the dimensions of the largest human cells (high-cost light sources are available, which offer resolutions in the region of 1 μm). This means that many biological samples have closely spaced interfaces that are difficult to resolve using standard OCT systems.

8.6.2   Examples of OCT Results

Examples of OCT images processed using the chromatic approach, described in Chapter 6, Section 6.2.3, are shown in Figure 8.6.1.

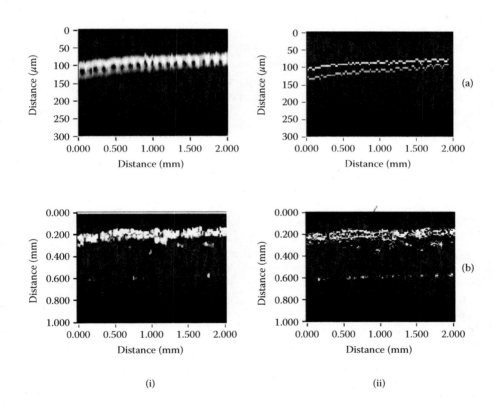

(i)   (ii)

FIGURE 8.6.1
Original (i) and chromatically processed (ii) OCT images. (a) air wedge between two glass interfaces with separation 17.5–30 microns, (b) onion slice with interfaces only resolvable in the processed image (indicated by arrows). (From Russell, C. D., Bullough, T. J., Jones, G. R., Krasner, N., and Bamford, K. J. (2003) In *Alt'03 International Conference on Advanced Laser Technologies: Biomedical Optics*, Vol. 5486 SPIE, Bellingham, WA, pp. 123–128. With permission.)

Figure 8.6.1a relates to an optically clean and well-defined double layer of reflecting interfaces, whereas Figure 8.6.1b relates to an optically turbid biological sample where interface definition is unclear. Figure 8.6.1a (i) shows an OCT image of an air wedge between two glass slides, with a constant increase in separation along the length of slides from ~17.5 to ~30 μm.

Figure 8.6.1a (ii) shows an image corresponding to that of Figure 8.6.1a (i) after chromatic processing as described in Chapter 6, Section 6.2.3. The distinction between the two interfaces is much better defined along the whole length of the wedge. The rate at which the wedge thickness decreases from 30 to 17.5 μm is clear. This shows the capability of the system to reduce background noise and highlight interfaces such that separations as low as at least 17.5 μm can be resolved.

Figure 8.6.1b (i) shows an unprocessed OCT image of an onion slice. The noise level present in this unprocessed image further limits the resolution to about 30 μm so that the fine structure of the top interface in Figure 8.6.1b (i) cannot be disclosed.

Figure 8.6.1b (ii) shows the chromatically processed form of Figure 8.6.1b (i). The resolution is improved to approximately 15 μm so that some of the fine details in the upper layer (arrows in Figure 8.6.1b (ii)) become visible.

The results thus demonstrate the capability of the chromatic processing to typically improve OCT image resolution in the presence of multilayers and turbidity. This means that interfaces separated by a distance less that the coherence length of the system can be resolved so that low-cost light sources could be used.

8.7   Summary

Chromatic techniques can be used for both *in vivo* and *in vitro* biological applications and with both optical fiber and remote sensing.

Optical reflectance, absorption, scattering, and interference effects from human tissue and blood can be processed chromatically.

With optical reflectance monitoring, the distributed nature of the spectra from biological tissue and fluids makes detection and identification of specific conditions difficult using conventional spectroscopic techniques, but this can be overcome with chromatic methods.

Chromatically based pulse oximetry can be made minimally sensitive to different skin tissue pigmentations of various ethnic groups and has a flexibility that enables oxyhemoglobin, carboxy-hemoglobin, and other hemoglobin derivatives to be distinguished.

Monitoring blood oxygenation of damaged tissue *in vivo* can be achieved using a vectorial combination of primary chromatic parameters that produce a monotonic variation with the degree of oxygenation.

The use of chromatic methods in conjunction with the ratio spectra has potential for identifying and distinguishing between different types of diseased tissues without the need for biopsy.

Chromatic processing of two-dimensional images can be used for distinguishing between different types of white blood cells *in vitro*. Significant distinction between various cells is possible in terms of chromatic H, which might be further enhanced by secondary chromatic processing (Chapter 1, Section 1.3.2 (d)).

Chromatic analysis can improve the resolution of low-coherence OCT images so that tissue interfaces separated by distances less than the coherence length of the system can be resolved.

References

Bedlow, A. J. (1995). Impact of skin cancer education on general practitioners' diagnostic skills. *Br. J. Dermatol.* 133(suppl 45): 29–30.

Borisova, E., Troyanova, P., and Avramov, L. (2005). Reflectance Measurements of Skin Lesions—Noninvasive Method for Diagnostic Evaluation of Pigmented Neoplasia. *Proc. SPIE Diagnostic Optical Spectroscopy in Biomedicine III* 5862: 20A1–20A11.

Dobreva, A. and Meshkov, T. (1984). *Atlas of Haematology.* Sofia, Bulgaria: Medicina and Fizcultura.

Emery, J. R. (1987). Skin pigmentation as an influence on the accuracy of pulse oximetry. *J. Perinatol* 7: 329–330.

Farina, B., Bartoli, C., Bono, A., Colombo, A., Lualdi, M., Tragni, G., and Marchesini, R. (2000). Multispectral imaging approach in the diagnosis of cutaneous melanoma: Potentiality and limits. *Phys. Med. Biol.* 45: 1243–1254.

Fercher, A.F., Depler, W., Hitzenberger, C.K., and Lasser, T. (2003). Optical coherence Tomography—Principals and Applications, *Rep. Prog. Phys.*, 66, 239–303.

Freeman, R. and Perks, D. (1990). Incidence and range of carboxyhemoglobin in blood for transfusion. *Anaesthesia Forum* 45: 583–585.

Gabrilczyk, M. R. and Buist, R. J. (1988). Pulse oximetry and post operative hypothermia: An evaluation of the Nellor N-100 in a cardiac surgical intensive care unit. *Anaesthesia* 43: 402–404.

Glomon, L. (2003). Source-Based Chromatic Methodology for optical fiber sensor systems. Ph.D. thesis, University of Liverpool.

Greenlee, R. T., Murray, T., and Bolden, S. et al. (2000). Cancer statistics—2000. *CA Cancer J. Clin.* 2000, 50: 7–33.

Griffiths, D. M., Isley, A. H., and Runciman, W. B. (1988). Pulse meters and pulse oximeters. *Anaesthesia and Intensive Care.* 16, No. 1, 49–53.

Jacques, S. L., Saidi, I. S., Ladner, A., and Oelberg, D. (1997). Developing an optical reflectance spectrometer to monitor bilirubinemia in neonates. SPIE Proceedings of Laser-Tissue Interaction VIII, ed. S.L. Jacques. 2975, 115–124.

Jones, G. R., Russell, P. C., Vourdas, A., Cosgrave, J., Stergioulas, L., and Haber, R. (2000). The Gabor transform basis of chromatic monitoring. *Meas. Sci. Tech.* (11), No. 5, 489–498.

Mackie, R. M. (1992). Clinical differential diagnosis of cutaneous malignant melanoma. In *Diagnosis and Management of Melanoma in Clinical Practice*, Ed. Kirkham. N. Springer, London.

Mackie, R. M., Hole, D., Hunter, J. A. A., Rankin, R. Evans, A. McLaren, K., Fallowfield, M., Hutcheon, A., and Morris, A. (1997). Cutaneous malignant melanoma in Scotland: Incidence, survival and mortality, 1979–94. *Br. Med. J.* 315: 1117–1121.

Marchesini, R., Brambilla, M., Clemente, C., Maniezzo, M., Sichirollo, A., Testori, A., Venturoli, D., and Cascinelli, N. (1991). *In vivo* spectrophotometric evaluation of neoplastic and nonneaoplastic skin pigmented lesions—1. Reflectance measurements, *Photochem. Photobiol.* 53: 77–84.

Marchesini, R., Cascinelli, N., Brambilla, M., Clemente, C., Masdheroni, L., Pignoli, E., Testori, A., and Venturoli, D. (1992). *In vivo* spectrophotometric evaluation of neoplastic and nonneoplastic skin pigmented lesions. II: Discriminant analysis between nevus and melanoma. *Photochem. Photobiol.* 55: 515–522.

Morton, C. A. and Mackie, R. M. (1998). Clinical accuracy of the diagnosis of cutaneous malignant melanoma. *Br. J. Dermatol.* 138: 283–287.

Nicholson, G. (2002). Private Communication. Centre for Disability Research and Innovation, Institute of Orthopaedics and Musculo-Skeletal Sciences, University College, London.

NIH (1992). Diagnosis and Treatment of Early Melanoma. NIH Ceonsens Statement. January 27–29 10: 1–26.

Pavlova, P. (2001). Creation of a pattern features model for identification in computer differential blood count. *Comp. Rend. de l'Acad. Bul. Sci* 54: 29–34.

Pavlova, P., Borisova, E., and Avramov. L. (2004). Automation of the Skin Cancer Diagnostics on the Bases of Reflectance Spectroscopy. *Proc. 9th Natl. Conf. Biomed. Phys. Eng.* 260–265.

Pavlova, P., Cyrrilov, K., and Moumdjiev, I. (1996). Application of HSV color system in identification by color of biological objects on the basis of microscopic images. *Medical Imaging and Graphics* 20: 357–364.

Pavlova, P. and Staneva, K. (2005). Dependence of the color in computer models on the color-reproducing signals discretization. *Elektrotechncka and Elektronica* 3–4: 34–37.

Rallis, I., Glomon, L., Deakin, A. G., Spencer, J. W., and Jones, G. R. (2005). Polychromatic monitoring of complex Biological systems, Proc. Complex Systems Monitoring Session of the International Complexity, Science and Society Conf. pp. 40–45 [Liverpool].

Ralston, A. C., Webb, R. K., and Runciman, W. B. (1991). Potential errors in pulse oximetry III: Effects of Interference, dyes, dyshemoglobins and other pigments. *Anaesthesia* 46: 291–295.

Rogers, D. (1985). *Procedural Elements for Computer Graphics.* McGraw-Hill, New York (*Transl., Mir, Moscow,* 1989).

Russell, C. D., Bullough, T. J., Jones, G. R., Krasner, N., and Bamford, K. J. (2004). Application of chromatic analysis for resolution improvement in optical coherence tomography (OCT), Proc. SPIE Vol. 5486, p. 123–128.

Russell, C. D., Deakin, A. G., and Jones, G. R. (2005). Chromatic Analysis for Signal Processing in Optical Coherence Tomography. *Proc. Complex Systems Monitoring Session, International Complexity, Science and Society Conference* (Liverpool), pp. 35–39.

Sahin, S., Rao, B., and Kopf, A. W. et al. (1997). Predicting ten-year survival of patients with primary cutaneous melanoma: Corroboration of a prognostic model. *Cancer* 80: 1426–1431.

Severinghaus, J. W. and Kelleher, J. F. (1992). Recent developments in pulse oximetry. *Anaesthesiology* 76: 1018–1038.

Van Assendelft, O. W. (1970). Spectrophotometry of Hemoglobin Derivatives. *Royal Van Gorcum,* Assen, Netherlands; and Charles C Tomas, Springfield, Ill.

Van Gemert, M. J. C., Jacques, S. L., Sterenborg, H. J. C. M., and Star, W. M. (1989). Skin optics. *IEEE Trans. Biomed Eng* 36: 1146–1154.

Wallace, V., Bamber, J., Crawford, D. Ott, R. and Mortimer, P. (2000b). Classification of reflectance spectra from pigmented skin lesions: A comparison of multivariate discriminant analysis and artificial neural networks. *Phys. Med. Biol.* 45: 2859–2871.

West, I. P. (1993). Pulse Oximetry for Magnetic Resonance Scanner Environments. Ph.D. thesis. University of Liverpool.

Wingo, P. A., Ries, L. A., and Giovino, G. A., et al. (1999). Annual report to the nation on the status of cancer, 1973–1996, with a special section on lung cancer and tobacco smoking. *J. Natl. Cancer Inst.* 91: 675–690.

Wukitsch, M. W., Petterson, M. T., Tobler, D. R., and Pologe, J. A. (1988). Pulse oximetry: Analysis of theory, technology, and practice, *J. Clin. Monit. Comput.*, 4, 4: 290–301.

9

Chromatic Methods Applied in Environmental Monitoring

A.G. Deakin and Y.R. Kolupula

CONTENTS

9.1   Introduction

Environmental monitoring embodies the need to address a variety of complex conditions. Chromatic techniques can be deployed in various ways for addressing such situations. A few examples to illustrate such deployments are presented. These include the monitoring of airborne, micron-sized particulates (PM10s), the compression and display of air quality data, the monitoring of conditions within anaerobic waste reduction facilities, the display of information indicating the progression of anaerobic reactions, the prognosis of outcomes, and the assessment of the growth of vegetation.

The fundamental bases for these various monitoring applications have been described in previous chapters. Monitoring scattered polychromatic light, which is used for determining airborne particulates, has been described in Chapter 7. Chromatic optical fiber and remote sensing, which are applied for monitoring anaerobic digestion, have been described in Chapter 6. The compression of discrete data sets, which has been used for displaying air-quality data and the progression of anaerobic digestion, has been described in Chapter 2, Sections 5 and 6. The approach for monitoring the growth of vegetation derives from a similar basis as that used for tissue monitoring described in Chapter 8.

9.2 Air-Quality Monitoring

9.2.1 Chromatic Monitoring of Airborne Particles Pollution

9.2.1.1 Introduction

The chromatic monitoring approach utilizing the accumulation of particles on microfiber filters (Chapter 7) can be deployed for monitoring airborne particulates in the size range 2.5–10 µm (PM 10) (Reichelt et al., 2006). The approach is not so much for high-precision measurements but to provide a means for economic wide-area monitoring for mapping particulates distributions (e.g., Chapter 2, Section 6) and for convenient indoor use. Both airborne particulates in outdoor urban areas and within smoky indoor environments have been addressed.

The technique may be deployed in the form of a portable, self-contained instrument unit (Figure 7.6, Chapter 7), or the accumulator unit may be addressed by a remote CCTV camera.

9.2.1.2 Portable Chromatic System

A portable chromatic system has been described in Chapter 7 that enables PM10 levels to be conveniently monitored at a number of different real-world sites such as along bus corridors, at bus termini, and within indoor environments.

Information about the distribution of airborne particulates along an urban/suburban bus corridor gathered from a vehicle traveling along the route and using the portable chromatic monitoring system has been presented in Chapter 2, Section 6. The way in which different forms of information could be extracted chromatically from the gathered data was described.

Particle concentration data gathered with the portable chromatic system (Figure 7.6, Chapter 7) at a city center bus terminus is shown in Figure 9.2.1. It shows PM10 concentrations relative to a reference level 30 µg/mL, as a function of the percentage of buses with particle traps on their exhausts multiplied

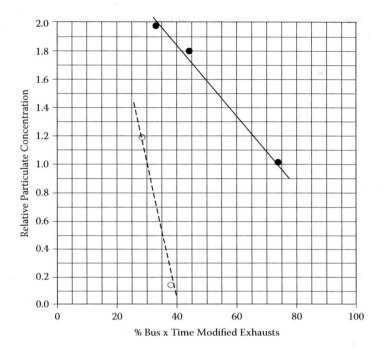

FIGURE 9.2.1
Variation of relative particle concentration with percentage of exhaust-modified buses × residence time (concentration normalized with respect to 30 mg/L). o—Without a high-polluting bus, •—with a high polluting bus. (From Jones, G. R., Spencer, J. W., Reichelt, T. E., Aceves-Fernandez, M., and Kolupula, Y. R. [2005]. Air Quality Monitoring in the City of Liverpool Using Chromatic Modulation Techniques. CATCH Final Project Report, University of Liverpool. With permission.)

by the residence time of the buses at the terminus (Jones et al., 2005). The data enables the effectiveness of installing exhaust particle traps on various fractions of the bus population using the terminus, to be quantified with regard to the improvement in airborne particulates. The deleterious effect of a single highly polluting bus (open circles, Figure 9.2.1) is noteworthy.

There is a reduction by 0.1 in the relative concentration of PM 10 particles per percent exhaust-modified bus time. However, if there is a heavily polluting bus amongst the cohort of buses, the reduction is less, being only 0.5 rather than 0.1.

An example of the kind of information that can be gathered within indoor environments with the portable unit is for monitoring particulates from tobacco smoke within public places such as restaurants. A comparison of particulate levels monitored with a portable system within a restaurant under conditions of unrestricted and restricted smoking suggests that there is almost an order of magnitude reduction in airborne particulates when smoking was restricted, compared to no restriction (Kolupula, 2007).

9.2.1.3   Remote CCTV Camera-Addressed Unit

Instead of the particle-accumulating filter being monitored by a miniature CCD camera attached to the particle filter unit as in the portable chromatic unit (Chapter 7, Figure 7.6), it may be addressed by a remote CCTV camera (Jones et al., 2005; Aceves-Ferenandez, 2005). The deployment of such a system for monitoring particulates across a busy city center thoroughfare using a CCTV camera from an already installed surveillance network is shown in Figure 9.2.2a. Figure 9.2.2b shows a typical zoomed view of the particle filter obtainable with such systems over distances of several meters. Sufficient resolution is available for clearly addressing the reference (R) and signal (S) zones.

The operation of such a system differs from that of the SCU system (Chapter 7) in that it is more susceptible to changes in ambient light conditions. These changes may be accommodated using the reference section of the filter unit (Figure 9.2.2b). Figure 9.2.3a shows the time variation of chromatic L for both reference and signal channels during the daylight hours over a four-day period close to a main city center thoroughfare. Both signal and reference channels exhibit diurnal variations and fluctuations, which are in phase. However, the signal channel also shows a deviation from the reference channel that increases with time and is due to the growth in accumulated particles.

Calibration curves of form shown in Figure 7.8b, Chapter 7, are employed to convert the L values into particle concentrations. However, in the case of remote monitoring, the L value of the reference channel at each time instant is employed to identify which is the relevant calibration curve at that instant. In this manner, the autogain of the camera (that cannot be disabled if the camera is also used for surveillance purpose, and which has similar effects to varying the intensity (voltage) of illumination source) can be accommodated.

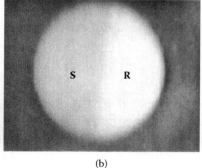

(a)   (b)

FIGURE 9.2.2
(See color insert following page 18). Remote CCTV camera monitoring of a particulates accumulator unit. (a) view of system, (b) typical zoomed view of particle filter (R—Reference, S—Signal). (From Jones, G. R., Spencer, J. W., Reichelt, T. E., Aceves-Fernandez, M., and Kolupula, Y. R. [2005]. Air Quality Monitoring in the City of Liverpool Using Chromatic Modulation Techniques. CATCH Final Project Report, University of Liverpool. With permission.)

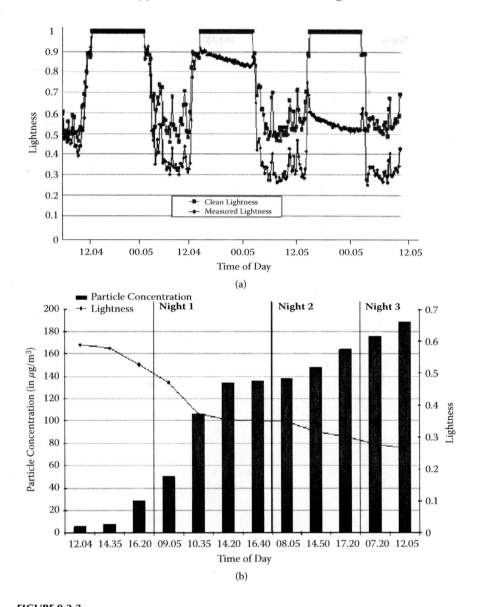

FIGURE 9.2.3
Time variation of L values from reference and signal channels, and particle concentration (accumulated): (a) signal and reference L variations, (b) signal L and accumulated particle concentrations. (From Jones, G. R., Spencer, J. W., Reichelt, T. E., Aceves-Fernandez, M., and Kolupula, Y. R. [2005]. Air Quality Monitoring in the City of Liverpool Using Chromatic Modulation Techniques. CATCH Final Project Report, University of Liverpool. With permission.)

FIGURE 9.2.4
(See color insert following page 18). Images of 10-µm particulates-loaded filter from a city center bus terminus showing changes in chromatic H. (Lamp voltage −3.5 V, filter–micro fiber 50). (From Jones, G. R., Spencer, J. W., Reichelt, T. E., Aceves-Fernandez, M., and Kolupula, Y. R. [2005]. Air Quality Monitoring in the City of Liverpool Using Chromatic Modulation Techniques. CATCH Final Project Report, University of Liverpool. With permission.)

The accumulation of particle concentration as a function of time for the 4-day duration of the test is shown in Figure 9.2.3b. The significance of this remote CCTV mode of operations is that it enables already installed area-wide CCTV surveillance networks to be utilized for obtaining distributions of PM10 particles across large urban areas in a cost-effective manner (Jones et al., 2005).

9.2.1.4   Particle Types

Different chromatic H signatures that did not conform to the H-L correlation locus of Figure 7.9b, Chapter 7, have been observed during airborne particles monitoring on some occasions at a busy city center bus terminus. Figure 9.2.4 shows a series of filter images having progressively different H values. These effects appear to be associated with particular vehicles and possibly their fuels.

Figure 9.2.5a, shows images of micropore filters that have been used to accumulate different kinds of PM10 particles. These are

- General airborne particles (Figure 9.2.5a (i))
- Tobacco smoke (Figure 9.2.5a (ii))
- Incense smoke (Figure 9.2.5a (iii))

The H-S chromatic signature of each sample is displayed on an H-S polar diagram (Figure 9.2.5b). This shows that the H-S chromatic signature of tobacco smoke is distinctly different from the other types of particles in being considerably more monochromatic (S = 0.9, compared with $S \cong 0.1$) and having a different dominant wavelength (H ~ 0°, compared with H ~ 66°). Thus, there is the potential for utilizing the chromatic H-S parameters for distinguishing between some particle types.

9.2.2   Air-Quality Data

Air-quality monitoring involves tracking levels of pollutants in the atmosphere. In addition to PM10-type particulates described in Section 9.2.1,

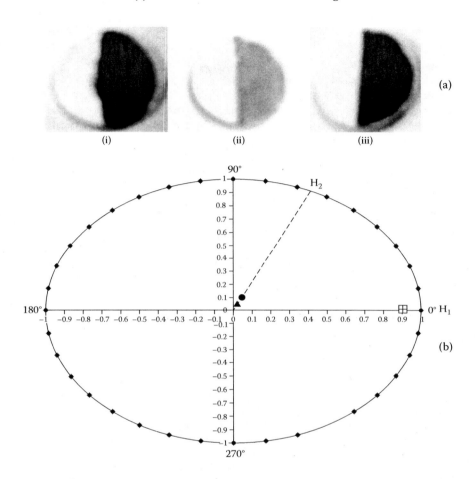

FIGURE 9.2.5
(See color insert following page 18). Chromaticity of different particle types accumulated on micropore filters: (a) images of signal and reference parts of filters—(i) urban air, (ii) incense smoke, (iii) tobacco smoke; (b) H–S polar diagram for different particles. ▲—urban air, ⊞—tobacco smoke, ●—incense smoke. (From Kolupula, Y. R. [2007]. Private Communication.)

traces of gases such as sulfur dioxide (SO_2), nitrogen dioxide (NO_2), and carbon dioxide (CO_2) are monitored at various sites at regular time intervals over prolonged periods. Consequently, a substantial amount of data is amassed that needs to be displayed in a compressed manner for ease of interpretation. Chromatic methods for addressing such large sets of discrete data have been described in Chapter 2, Sections 5 and 6. The approach can be used for addressing the problem of displaying and interpreting the air-quality data.

By way of an illustration of the deployment of the chromatic approach for interpreting air-quality data, a set of three main constituent components of interest may be selected. These are SO_2, NO_2, and PM10. These components were monitored at hourly intervals for a 6-month period at an urban site in the United Kingdom. The levels of the three pollutants are treated as three chromatic processor outputs $R(=SO_2)$, $G(=NO_2)$, and $B(=PM10)$, which are normalized with respect to average, expected levels. They are then transformed into chromatic H,L,S parameters that can be displayed on polar diagrams; the H-L diagram being of particular interest. This diagram enables the relative balance of the three pollutants to be observed over the time period of the monitoring.

Figure 9.2.6 shows such an H-L polar diagram for the data collected over the 6-month period. The H parameter represents the type of pure pollutant (SO_2—0, NO_2—120, PM10—240) or combinations of pollutants. The L parameter represents relative levels of pollutants, with the inner circle representing the mean pollutant level and the three outer circles indicating 1, 2, 3 standard deviations from the mean, respectively. The color of each data point indicates

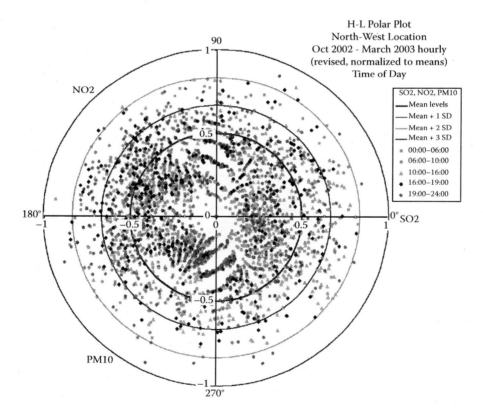

FIGURE 9.2.6
(See color insert following page 18). H-L polar diagram of SO_2, NO_2, PM10 airborne pollutants measured hourly over a six-month period, U.K. urban area. (H—pollutant; color of points— time of day; concentric circles (L)—mean level plus 1, 2, 3 standard deviations). (From Deakin, A. G., Rallis, I., Zhang, J., Spencer, J. W., and Jones, G. R. [2005]. Towards holistic chromatic intelligent monitoring of complex systems. *Proc. Complex Systems Monitoring Session, International Complexity, Science and Society Conference* (Liverpool). pp. 16–23. With permission.)

the time of day for each measurement: green (00.00–06.00), orange (06.00–10.00), light blue (10.00–16.00), dark blue (16.00–19.00), and brown (19.00–24.00).

Displaying the data in this manner enables the time of day to be identified when the pollutant levels exceed the mean levels as well as indicating the dominant constituent pollutants during each time period. For example, there are six occasions when the mean +2 standard deviations was exceeded by the PM10 pollutant with secondary SO_2 or (H~ 200–300°, Figure 9.2.6) during the time period 19.00–24.00 (brown points). This may suggest that evening traffic could be a contributory cause.

As another example, there appears to be a predominance of PM10/SO_2 levels (H~ 270–360°) up to the +2 standard deviation during the midday period 10.00–16.00 h (light blue points).

The same data may be conveniently reorganized computationally along different dimensions, e.g., day of the week, month of the year (Chapter 2, Section 6), so that further patterns may be revealed.

9.3 Anaerobic Waste Treatment

9.3.1 Introduction

Anaerobic digestion (AD) (Hobson et al., 1984; Williams, 1998; Tonge et al., 2003) is an approach for generating various gases from waste organic matter using naturally existing microbiological processes of fermentation and digestion. Because energy can be recovered in a carbon-neutral manner by AD in the form of methane (natural gas) along with other valuable gaseous, liquid, and solid by-products (e.g., Hobson et al., 1984), it is being increasingly investigated for treating biodegradable waste on industrial scales, for example, by food processing industries.

For such purposes, conditions within the organic waste need to be optimized for encouraging the growth of the required microbial colonies for promoting the progression of biodigestion. Two indicators of the conditions needed for such promotion are temperature and acidity (pH level) of the leachate that is being produced. The coloration of the leachate can also be a useful indication of the progression of the biodegradation process.

Because methane and hydrogen are produced during the degradation process, conditions within the waste-processing cells are potentially explosive, and so the cells are classified as zone zero-hazard environments. The use of electronic sensors and instruments within such environments is prohibited unless special explosion-proof assemblies are utilized. The use of optical fiber and remote sensing offers attractive alternative sensing possibilities because of their reduced susceptibilities to induce explosions.

An experimental, industrial-scale, anaerobic digester unit on which chromatic sensing and monitoring has been demonstrated consisted of 5×64 m^3, air-sealed cells for containing organic waste plus an integral greenhouse for

FIGURE 9.3.1
(See color insert following page 18). Experimental, industrial-scale anaerobic waste treatment unit: (a) view of the waste cells—greenhouse integrated unit, (b) vertical section of the unit showing the relative locations of the waste-processing cells and greenhouse, (c) filling a treatment cell with organic waste. (Courtesy AMEC. With permission.)

cultivating plants with the by-products (heat, carbon dioxide, etc.) from the digester (Figure 9.3.1a) and an electric power generator. The five cells were interconnected by leachate recirculating pipes (Figure 9.3.2a), which carried instrumentation for monitoring each cell's interconnection with the pipe system. During operation, the temperature and acidity (pH) of the leachate need to be optimized within tolerances during operation to optimize the anaerobic reactions and thus maximize the production of methane (for electric power generation) and other by-products (e.g., heat, CO_2 for use in the greenhouse). A typical record of the biogas and methane outputs as a function of time during a waste-reduction cycle is shown in Figure 9.3.3 along with the temperature and pH measured outside the cell on the interconnecting pipe-work (Figure 9.3.2a). Fluctuations in biogas production occur, synchronized with variations in temperature and pH levels outside the required operational regime.

The operation of such a facility is subject to many hazards, for example, the potentially explosive nature of the methane-containing atmosphere in the digester cells, the chemically aggressive nature of the leachate, etc. Sensing and monitoring the leachate condition, which has a complex composition, is made more difficult by the need to respect the potentially explosive and chemically aggressive nature of the conditions.

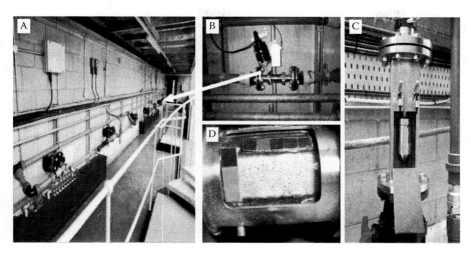

FIGURE 9.3.2
(See color insert following page 18). Chromatic monitoring instrumentation at the anaerobic treatment facility: (a) pipe-work for recycling digestate, (b) CCTV remote monitor of leachate, (c) chromatic optical fiber pH and temperature sensors lance (d) view of transparent pipe section and thermochromic elements. (Courtesy AMEC. With permission.)

9.3.2 Chromatic Sensing and Monitoring Techniques

Monitoring anaerobic digestion conditions with chromatic techniques has involved three different aspects. These are

1. Sensing the temperature and pH levels of the leachate with chromatic optical fiber sensors within the potentially explosive environment of a cell.

2. Monitoring the temperature and condition of the leachate and the biogas volume in the recirculating pipes with remote chromatic sensing.

3. Displaying patterns of the progression of the production of biogas components during an anaerobic processing cycle with chromatic polar diagrams for prognosis purposes.

9.3.2.1 Chromatic Optical Fiber Sensing

Optical fiber sensing provides electrical isolation for the sensing element inserted into potentially explosive environments because of the nonelectrical nature of the optical fibers and sensors. They can also be assembled from materials that can withstand many chemically aggressive environments. For leachate monitoring, the chromatic optical fiber sensors were housed within lances (Figure 9.3.2b), which could be inserted into the 4-m-deep cells for monitoring the temperature and pH at different levels and with optical fiber transmission over distances of 30 m away from the hazardous area.

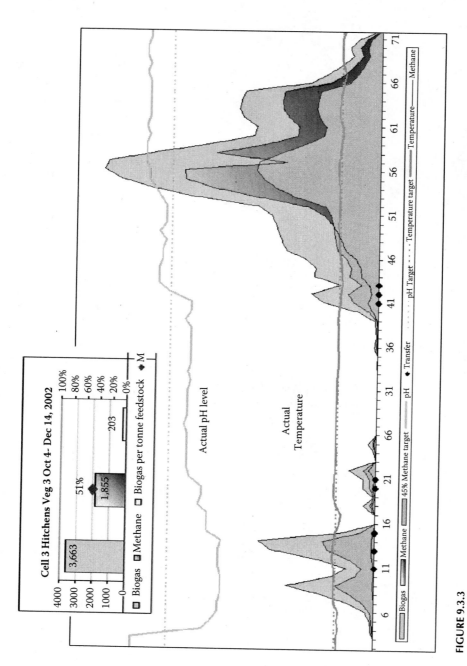

FIGURE 9.3.3
Time variation of digestate temperature and pH, and of biogas production. (Courtesy AMEC. With permission.)

The temperature-sensing element was a cobalt chloride solution unit (Chapter 5, Section 5.2.4.3). The pH-sensing element was a color-varying, chemical indicator (Chapter 5, Section 5.2.4.2). Both optical elements were addressed chromatically using distimulus chromatic processing (Chapter 3, Section 3.3.2b). Typical calibration curves of chromatic H versus temperature, and of chromatic H versus pH level, have been given in Chapter 5, Section 5.2.4 (Figures 5.2.4.2b and 5.2.4.3). Measurements within a digester cell showed diurnal variations in both temperature and pH (Figure 9.3.4a and b).

The monitoring was undertaken via 30-m-long optical fiber links. The pH indicator element remained operational for 3 months without performance degradation due to ingress of particulates or bleaching of the indicator dye. The chromatic pH sensor results agreed well with more limited, periodic pH measurements made with a handheld electronic unit on samples of the leachate extracted from the cell (Figure 9.3.4b).

9.3.2.2 Remote Chromatic Monitoring

Remote chromatic monitoring of the leachate involved addressing an illuminated section of transparent pipe inserted into the recirculating pipe with a CCTV camera (Figure 9.3.2a, and d). The transparent pipe section also carried a series of thermochromic elements (horizontal row, Figure 9.3.2d; Figure 6.3.4.4, Chapter 6) and a number of chromatic reference elements (vertical row) to compensate for any illumination variation. The images captured with the CCTV camera enabled three condition indicators to be monitored from each CCTV frame captured:

1. The temperature of the glass wall confining the leachate could be determined from the thermochromic elements (Chapter 6, Section 6.3.4.2).

2. An indication of the leachate condition could be given by the chromaticity of the leachate itself.

3. The amount of biogas formed within the clear glass section of the pipe (due to the continuing microbial activity in the leachate) can be determined by applying space chromatic domain filters (Chapter 2, Section 4) to the pipe section (Rallis, 2004).

The time variation of the leachate temperature and the wavelength chromaticity are shown in Figure 9.3.5a, covering a period of 2 weeks. The time variation of the hue of the leachate at two locations is similar, indicating the extent of uniformity of conditions.

Gas formation within the transparent pipe section is apparent in the form of bubbles on the frames given in Figure 9.3.5b. The frames cover the time period from 19.30 to 03.30 h and illustrate how the size of the gas bubbles increases with time. When processed with space-domain chromaticity (Chapter 2, Section 4) (Rallis, 2004), the volume of biogas produced can be

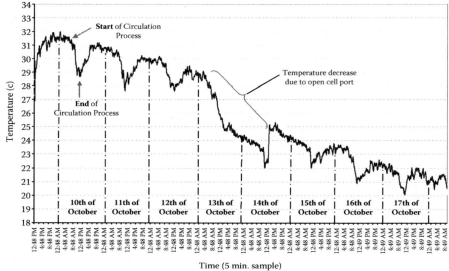

(a)

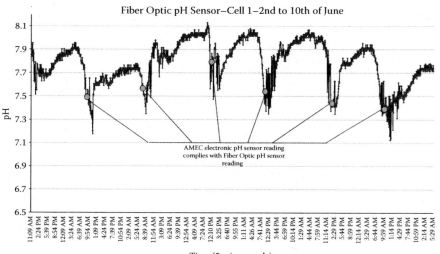

(b)

FIGURE 9.3.4
Chromatic optical fiber sensing of pH and temperature during a waste-reduction cycle: (a) diurnal variation of temperature, (b) diurnal variation of pH. (From Rallis, I., Deakin, A. G., Spencer, J. W., and Jones, G. R. [2005a]. Novel sensing techniques for industrial scale bio-digesters. 17th Int. Conference on optical fiber sensors. *Proc SPIE* 5855 pp. 110–113. With permission.)

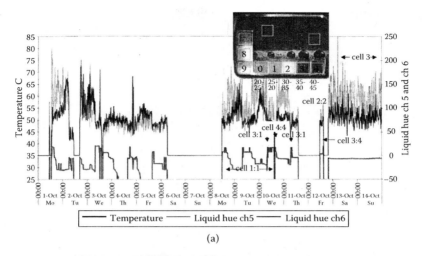

(a)

(b)

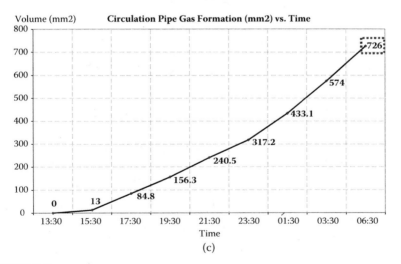

(c)

FIGURE 9.3.5
(See color insert following page 18). Time variation of conditions within the transparent length of the recirculating pipe: (a) time variation of leachate temperature and chromaticity, (b) CCTV images of the pipe at various times showing biogas bubbles, (c) time variation of the volume of biogas bubbles determined using space chromatic processing of the CCTV images. ((b), (c) Rallis, I. (2004) Intelligent Chromatic Fiber Optic Sensors and Monitoring Systems for Enhancing Useful By-Products from Anaerobic Digestion. Ph.D. thesis, University of Liverpool; (a) Courtesy AMEC; Johnson, M. S. (2004) Research into the Anaerobic Biodigestion of Wastes, University of Liverpool Report.)

determined, leading to the time variation curve shown in Figure 9.3.5c. This shows that a total volume of 726 mm^3 is produced from the leachate in the length of pipe observable through the glass window and within the time interval 13.30 to 06.30 h.

9.3.2.3 Progression of the Anaerobic Cycle

The progression of an anaerobic cycle may be determined from the proportions of various gases produced by the process at each time instance, the composition throughout the entire cycle being well documented scientifically (Williams, 1998; Farquhar and Rovers, 1973). The relevant gaseous components are N_2, O_2, Ar, H_2O, CO_2, H_2, and CH_4. An assembly of the levels of each component can be addressed by discrete chromatic processing, as described in Chapter 2, Sections 5 and 6. The ordered list of gases and their levels, commencing with air (N_2, O_2) and ending with methane (CH_4) are addressed by three triangular chromatic processors (R, G, B) whose outputs are transformed into chromatic H, L, S values (Equations 1.2–1.7, Chapter 1).

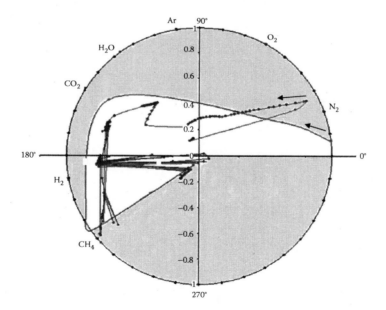

FIGURE 9.3.6

H-S chromatic polar diagram for methanogenesis. (From Zhang, J., Du, X., Yuan. W. D., Deakin. A., Spencer. J. W., Jones. G. R., Gibson, J. R., Hall. W. B., McGrail, A. A., and Tonge. H. [2001]. Tracking trends in the chemical composition of systems using chromatic mapping. *Proc. IEE Seminar on Intelligent and Self Validating Instruments* (Sensors and Actuators). (2001), London. With permission.)

A polar diagram of H versus S is then produced, where H represents the dominant member of the gas mixture and S the effective spread of components at that time instant (Figure 9.3.6). The order of the gaseous components appear along the azimuthal axis H (N_2–CH_4). Taking scientifically well-established gas composition data (Williams, 1998; Farquhar and Rovers, 1973) for various stages of a digestion cycle, a calibration curve in H-S space can be established (boundary between shaded and white regions Figure 9.3.6).

Chromatically transformed gas composition data obtained at various times during a digestion cycle can be registered on the H-S polar diagram (continuous curve, Figure 9.3.6) and compared with the calibration curve to assess the progress of the waste-reduction cycle. If the monitored H, S values deviate from the calibration curve, then corrective action can be taken by adjusting the leachate temperature and pH levels through the recirculation of leachate between cells. The example of the progression of an anaerobic cycle (continuous curve) given in Figure 9.3.6 shows substantial short-term deviations from the calibration curve, which, in this instance, were due to the cell being opened for visual inspection.

9.4   Vegetation Growth

The availability of white-light-emitting diodes (LEDs) with increased opto-electronic efficiency and reduced heat production than conventional illumination sources (e.g., filament lamps, energy-saving lamps, etc.) provides new opportunities for assisting photosynthesis in vegetation. This can be of value, for example, in greenhouse atmospheres enriched with carbon dioxide (for enhancing plant growth) plus heating provided as a by-product of anaerobic digestion of biological waste (Tonge et al., 2003). Establishing the effectiveness of such possibilities requires convenient on-line monitoring, which can involve chromatic techniques.

A schematic representation of a laboratory system for investigating such possibilities is shown in Figure 9.4.1a. This illustrates how test plants (cress) can be exposed to white LED light (regions 2, 4 Figure 9.4.1b), whereas others are not exposed (regions 1, 5) or only partly exposed (region 3). The figure also shows how the plants in each region are chromatically monitored with a CCTV camera connected to a computer for capturing and processing R, G, B signals (Rallis et al., 2005).

Figure 9.4.1b (i) shows an image of the plants during growth as viewed by the camera and with the LED-exposed regions designated by the red circles. (Also shown are three elements of a thermochromic temperature sensor (Chapter 6, Section 6.3.4.2), which allowed the temperature to be monitored chromatically and simultaneously with the CCTV camera).

Figure 9.4.1b (ii) shows a side view of the growing plants, such images being recorded at regular time intervals. The LED-exposed regions are again

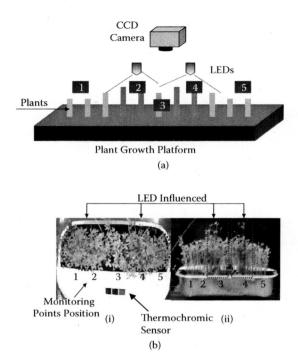

FIGURE 9.4.1
Chromatic monitoring of LED-assisted vegetation growth: (a) layout of LED-illuminated system, (b) view of vegetation: (i) plan view, (ii) side view. (From Rallis, I., Glomon, L., Deakin, A. G., Spencer, J. W., and Jones, G. R. [2005b]. Polychromatic monitoring of complex biological factors. *Proc. Complex Systems Monitoring Session*, Int. Complexity, Science and Society Conference (Liverpool), pp. 40–45. With permission.)

indicated by the red rectangles. Visual inspection of such images indicated that within the LED-illuminated regions there was

1. More pronounced growth (Figure 9.4.1b (ii))
2. A deeper coloration of the plants (Figure 9.4.1b (i))

The results of such tests show that the height of the LED-exposed plants was about 45% greater than that of the plants not exposed (Figure 9.4.2a). Also, the LED-exposed plants had higher H and S values (suggesting that they were more monochromatic in the green spectral region) but lower values of lightness (consistent with part of the spectrum being absorbed/scattered) (Figure 9.4.2b).

The H, L, S values for each differently illuminated condition may be combined vectorially to yield a chromatic vector C_T (Equation 1.13, Chapter 1)

$$C_T = \sqrt{a_H \cdot H_N^2 + a_L\,(b_L - L_N)^2 + a_S\,(b_S - S_N)^2} \qquad (9.4.1)$$

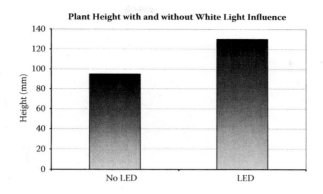

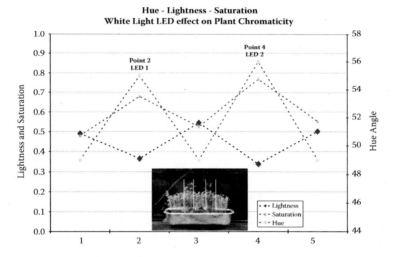

FIGURE 9.4.2
Sample test results for vegetation exposed to different illumination: (a) comparison of vegetation heights, (b) H, L, S parameter values for the vegetation foliage as a function of position. (From Rallis, I., Glomon, L., Deakin, A. G., Spencer, J. W., and Jones, G. R. [2005b]. Polychromatic monitoring of complex biological factors. *Proc. Complex Systems Monitoring Session*, Int. Complexity, Science and Society Conference (Liverpool), pp. 40–45. With permission.)

The results of Figure 9.4.2b indicate that dH/dL is negative, so that $b_L = 1$ (Chapter 1, Equation 1.15), and that dH/dS is positive, so that $b_S = 0$ (Equation 1.16). This then allows the correlation between plant height and the chromaticity of the foliage to be examined on a graph of C_T against plant height (Figure 9.4.3). The result shows that both the plant height and chromatic parameter C_T are greater for the LED-exposed specimens compared with those exposed to the general laboratory fluorescent light alone.

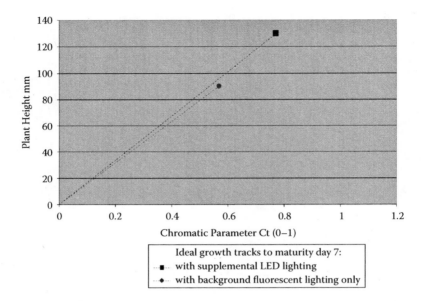

FIGURE 9.4.3
Chromatic parameter C_T of foliage as a function of vegetation height. (From Rallis, I., Glomon, L., Deakin, A. G., Spencer, J. W., and Jones, G. R. [2005b]. Polychromatic monitoring of complex biological factors. *Proc. Complex Systems Monitoring Session*, Int. Complexity, Science and Society Conference (Liverpool), pp. 40–45. With permission.)

9.5 Summary

The chromatic approach in combination with a micropore particle filter provides a flexible means for monitoring airborne particles in a robust manner and at various kinds of sites.

The methodology may be deployed in a variety of systems such as a portable unit or as part of a CCTV area network.

Compensation for system drift and changes in extraneous light that can occur with remote sensing can be accommodated.

Discrimination between some categories of particles (e.g., tobacco smoke and vehicle exhaust particles) seems feasible.

Indications of deviations from the *normal* airborne particulate population (e.g., different exhaust emissions) may be possible.

Air pollution data sets can be chromatically manipulated to produce polar diagrams on which patterns of different trends of excessive pollutant levels can be conveniently identified.

Chromatic optical fiber sensors for monitoring temperature and acidity (pH) are sufficiently robust for deployment within the chemically aggressive and potentially explosive environment of anaerobic digester cells to provide detailed indications of how optimum the conditions are for producing efficient digestion.

Remote chromatic monitoring of the circulated digestate is capable of providing nonintrusively, indicators of the digestate conditions and biogas production.

The progression of an anaerobic digestion cycle may be displayed on a chromatic H-S polar diagram for assisting prognosis of gas production, etc.

Monitoring the growth of at least one type of vegetation (cress) appears to be viable using a composite chromatic parameter C_T.

References

Aceves-Fernandez, M. A. (2005). Chromatic Intelligent Systems for Pollution Monitoring. Ph.D. thesis, University of Liverpool.

Deakin, A. G., Rallis, I., Zhang, J., Spencer, J. W., and Jones, G. R. (2005). Towards holistic chromatic intelligent monitoring of complex systems. *Proc. Complex Systems Monitoring Session, International Complexity, Science and Society Conference* (Liverpool), pp. 16–23.

Farquhar, G. J. and Rovers, F. A. (1973). Gas production during refuse decomposition. *Water, Air, Soil Pollut.* 2, 483–495.

Hobson, P. N., Reid, W. G., and Sharma, V. K. (1984). Anaerobic conversion of agricultural waste to chemicals or gases. In *Anaerobic Digestion and Carbohydrate Hydrolysis of Waste* (Ed. Ferrero, C. L. et al.) Elsevier, New York.

Johnson, M. S. (2004). Research into the Anaerobic Biodigestion of Wastes, University of Liverpool Report.

Jones, G. R., Spencer, J. W., Reichelt, T. E., Aceves-Fernandez, M., and Kolupula, Y. R. (2005). Air Quality Monitoring in the City of Liverpool Using Chromatic Modulation Techniques. CATCH Final Project Report, University of Liverpool.

Kolupula, Y. R. (2007). Private Communication.

Rallis, I. (2004). Intelligent Chromatic Fiber Optic Sensors and Monitoring Systems for Enhancing Useful By-Products from Anaerobic Digestion. Ph.D. thesis, University of Liverpool.

Rallis, I., Deakin, A. G., Spencer, J. W., and Jones, G. R., (2005a). Novel sensing techniques for industrial scale bio-digesters. 17th Int. Conference on optical fiber sensors. *Proc SPIE* 5855, pp. 110–113.

Rallis, I., Glomon, L., Deakin, A. G., Spencer, J. W., and Jones, G. R. (2005b). Polychromatic monitoring of complex biological factors. *Proc. Complex Systems Monitoring Session*, Int. Complexity, Science and Society Conference (Liverpool), pp. 40–45.

Reichelt, T. E., Acevez-Fernandez, M. A., Kolupula, Y.R., et al. (2006). Chromatic modulation monitoring of airborne particulates. *Meas. Sci. Technol.* 17, pp. 675–683.

Tonge, H., Deakin, A. G., and Yan, J. D. (2003). Anaerobic digestion in waste reduction—the challenge. In *Recycling and Reuse of Waste Materials, Advances in Waste Management and Recycling* (Eds. Dhir, R. K., Newlands, M. D., and Halliday, J. E.), University of Dundee, September 9–11, 2003, pp. 213–222.

Williams, P. T. (1998). The main steps in anaerobic digestion of municipal waste. In *Waste Treatment and Disposal,* John wiley and Sons Ltd., London, (2001), London, p. 392.

Zhang, J., Du, X., Yuan. W. D., Deakin. A., Spencer. J. W., Jones. G. R., Gibson, J. R., Hall. W. B., McGrail. A. A., and Tonge. H. Tracking Trends in the Chemical Composition of Systems using Chromatic Mapping. *Proc. IEEE Seminar on Intelligent and Self Validating Instruments* (Sensors and Actuators) John Wiley and Sons Ltd., London, (2001), London.

10

Chromaticity of Acoustical and Vibration Signals

A.G. Deakin and X. Zhang

CONTENTS

10.1   Introduction

Acoustical signals carry energy and information about their source, which is propagated as sound waves traveling through a physical medium in the form of gas, solid, or liquid. The information carried as acoustical waves exists in two orthogonal domains—time and frequency (Jones et al., 2000). Frequency ranges of interest acoustically can extend from a few Hz to several hundred kHz. Some information is carried in terms of the combination of frequencies, their amplitude, and phases.

As in the case of light waves and optical spectra, the information carried by sound waves may be either at a single frequency (cf. monochromatic, Chapter 1, Section 1.3.1.1) or a complex mixture of frequencies (cf. polychromatic). It is in the unravelling of information contained in such acoustical signals that the chromatic approach is useful. In the time domain, information is contained in the time-varying behavior of the signal, consisting of its profile, position, and gradient, all changing in time, and which may vary over different time scales.

This chapter, therefore, addresses situations where acoustical signals are generated under complex conditions, and information about the behavior of their source is contained in complex ways in the time and frequency domains. The signals are captured over broad frequency ranges by general sensors (e.g., microphones or vibration sensitive optical fibers), which are cost-effective to deploy, rather than being acquired by dedicated, specialized narrow-band sensors.

The acoustical signals of interest in monitoring are often of a pulsatile nature. Both continuous and pulsatile or impulsive signals can be chromatically processed. Chromatic processing is usefully applied for deriving primary chromatic parameters from the complex signal in the time domain and from the complex spectra in the frequency domain. As in the optical domain, it characterizes the information in terms of hue, saturation, and lightness. In the frequency domain, H, S, L represent, respectively, dominant frequency, bandwidth, and effective signal strength. In the time domain they represent dominant time period, period extent, and effective energy.

Use can also be made of second-generation (hierarchical) chromatic processing (Chapter 1, Section 1.3.1.2 (d)) for identifying emergent features embedded in the acoustical signals. Therefore, there are several options available for processing acoustical signals chromatically. These are discussed in Section 10.2, followed by examples of signals from real situations.

10.2   Choice of Processing Procedures

There are several options for processing acoustical signals chromatically as indicated in Figure 10.2.1. Various forms of a signal and processor deployment at different stages of the procedure are shown in Figure 10.2.2.

The first step is to threshold the time varying signal as it is captured in order to remove excessive amplitude artifacts (Figure 10.2.1 (i)). Thereafter there is a major choice between whether the signal is to be sampled by nonorthogonal processors in the time or frequency domains.

With frequency-domain chromaticity, there is a choice as to whether the signal is initially processed in hardware (through the deployment of electronic filters) or in software (through the deployment of FIR filters or Fast Fourier Transform after digitization, e.g., Figure 10.2.1 (ii)). The frequency range may be subdivided to segregate some information domains (Figure 10.2.1

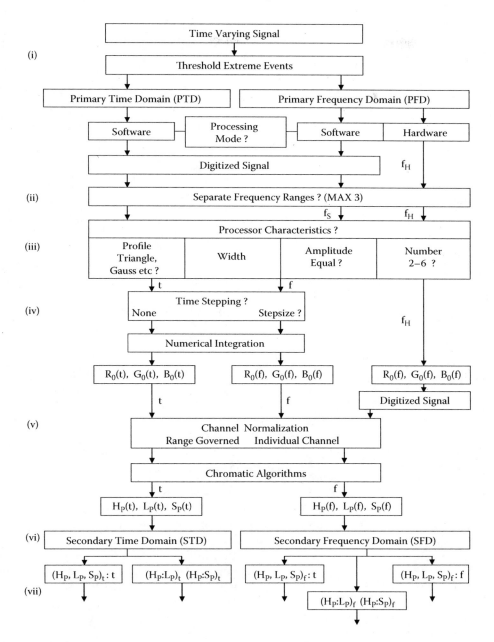

FIGURE 10.2.1
Decision diagram for choice of chromatic procedures ((i)–(vii) reference to Figure 10.2.2).

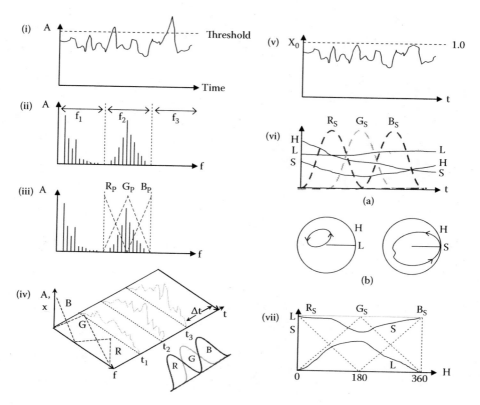

FIGURE 10.2.2
Schematic diagrams of a signal at various stages of processing ((i)–(vii) reference to Figure 10.2.1).

(ii)), e.g., avoiding well-defined electrical interference. In practice, no more than three separate regions have been found to be necessary. The characteristics of the processors need to be chosen (Chapter 2, Section 2) with respect to profile, width, amplitude, and number of processors in the range 2–6 (Chapter 2, Section 3; Figure 10.2.1 (iii)). For the software approach, the frequency-domain processors may be stepped in time so that the step size (Δt Figure 10.2.2 (iv)) needs to be defined. The output of each processor ($R_0(f)$, $G_0(f)$, $B_0(f)$) is obtained by numerical integration in the software case and directly from the processors in the hardware case, these outputs being digitized for further manipulation. $R_0(f)$, $G_0(f)$, $B_0(f)$ may be normalized with respect to a nominal range value common to each. For retrospective processing, normalization of each channel separately with respect to its maximum recorded value (Figure 10.2.1 (v)) can provide additional discrimination possibilities.

The primary frequency-domain chromatic parameters ($H_p(f)$, $L_p(f)$, $S_p(f)$) or ($x_p(f)$, $y_p(f)$, $z_p(f)$) are evaluated from $R_0(f)$, $G_0(f)$, $B_0(f)$ with the algorithms

(Chapter 1, Equations 1.2–1.7). These chromatic parameters may be presented as time-varying graphs or on chromatic polar diagrams (Figure 10.2.2 (vi) (a), (b)).

With time domain chromaticity, the procedure is similar to that for the frequency-domain software approach apart from the time stepping, if used, occurring in the same domain as that in which the filters are deployed (Figure 10.2.1 (iv)).

Once the primary domain chromatic parameters in either time ($H_p(t)$, $L_p(t)$, $S_p(t)$) (PTD) or frequency (PFD) ($H_p(f)$, $L_p(f)$, $S_p(f)$) have been determined, secondary domain parameters may be calculated. The secondary processing involves addressing the variation of the primary domain chromatic parameters (H_p, L_p, S_p) with chromatic processors (R_s, G_s, B_s) (Figure 10.2.2 (vi) (a)). The secondary processing may be undertaken with respect to time for the PTD parameters, and with respect to time or frequency for the PFD parameters.

Often H-L, H-S chromatic polar diagrams exhibit loci for various monitored conditions that have well defined, different but complicated, patterns. Although such patterns are qualitatively different, the differences need quantification in order to provide monitoring metrification. This may be achieved through the secondary processing of the H-L, H-S curves (Figure 10.2.2 (vii)), whereby the pattern is quantified by the coordinates H_{LS}, L_{LS}, S_{LS}, H_{SS}, L_{SS}, S_{SS} (cf. convoluted arc columns, Chapter 4, Section 4.3.3). For H-L, H-S patterns involving several convolutes, the dependent variables may be changed from $L(H)$, $S(H)$ to $\Sigma_H L(H)$, $\Sigma_H S(H)$.

10.2.1 Software and Hardware Chromatic Filters for Frequency-Domain Acoustic Signals

Frequency-domain processing of acoustical signals may be performed either in software (with previously sampled time domain signals) or with hardware filters (incorporated directly into the acoustic detection circuitry) (Russell et al., 1998).

In the software case, the signal is digitized, filtered, and Fast Fourier Transformed into the frequency domain before being addressed with three non-orthogonal filters. Each filter output is obtained by integrating the product of the filter response and signal amplitude at each frequency (Appendix 1.I, Chapter 1) before transformation into chromatic parameters x, y, z; H, L, S, etc. (Equations 1.2–1.7, Chapter 1). The method is useful for analyzing data following a nonrepetitive event but is computationally intensive.

In the hardware case with the filters incorporated into the detection circuitry, the computational requirements are significantly reduced because the need to capture considerable amounts of data to perform the Fourier transform and to integrate with respect to the chromatic processors' responses is removed.

Comparison of the two techniques can be made by chromatically analyzing known frequency signals (e.g., 1–5 kHz range in a series of 1-kHz steps) from an acoustical source with each technique (Russell et al., 1998). Approximately equivalent chromatic filter responses in software (Figure 10.2.3a) and hardware (Figure 10.2.3b) are needed for making such a comparison (e.g., covering a frequency range 2–8 kHz).

Values of the chromatic parameters L, H, S derived with equations 1.2–1.7, Chapter 1, with each technique are compared in Figure 10.2.4a, b, c, respectively. The variation of H and S with frequency shows similar trends (although not identical due to detailed differences in the software and hardware filter

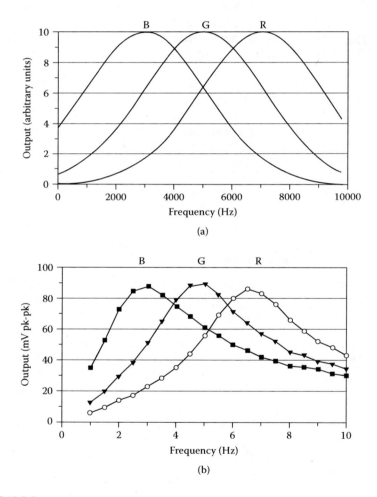

(a)

(b)

FIGURE 10.2.3
Responsivities of acoustic filters as a function of frequency: (a) Gaussian software filters, (b) Measured electronic filters (Peak R–7 kHz, Peak G–5 kHz, Peak B–3 kHz). (From Russell, P. C., Cosgrave, J., Tomtsis, D., Vourdas, A., Stergioulas, L., and Jones, G. R. (1998). Extraction of information from acoustic vibration signals using Gabor transform type devices. *Meas. Sci. Technol.*, 9, pp. 1282–1290. With permission.)

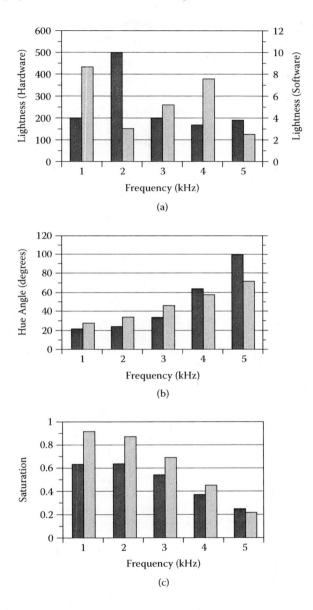

FIGURE 10.2.4
Comparison of chromatic parameters determined with software and hardware filters. (a) *L*, (b) *H*, (c) *S* (mean of 5 tests). (Russell, P. C., Cosgrave, J., Tomtsis, D., Vourdas, A., Stergioulas, L., and Jones, G. R. (1998). Extraction of information from acoustic vibration signals using Gabor transform type devices. *Meas. Sci. Technol.*, 9, pp. 1282–1290. With permission.)

responses (Figure 10.2.3)). The L values derived with each technique are similar but not identical due to a combination of detailed differences in the filter responsivities and reproducibility of the signal source from test to test. Thus, the two techniques give similar results indicating the validity of both approaches. However, test-to-test repeatability assessments indicate the software approach to be inferior to the hardware approach, with up to 10% variation in chromatic parameter values at the spectral extremities compared to less than 1% with the hardware technique (Figure 10.2.4). The inferior performance of the software technique is ascribed to truncation errors.

10.3   Pulsatile Nature of Acoustical Signals

Pulsatile or impulsive signals (e.g., the rapid opening of a mechanical switch, a road vehicle driving over a rut) arise from sporadic events. They may be single isolated pulses or repetitive pulses (e.g., reciprocating machinery), and

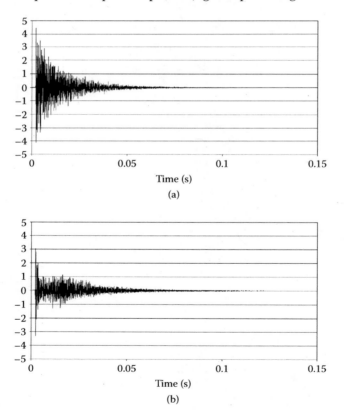

FIGURE 10.3.1
Acoustical signals obtained by mechanically applying an impulse to industrial rollbars: (a) a good-quality rollbar, (b) a bad-quality rollbar.

the latter may be distinct or overlapping to form quasi-continuous signals. In either case, departure of the signal characteristics from a preestablished norm can be indicative of a fault developing.

Acoustical/mechanical vibration signals produced for monitoring purposes are often of a pulsatile form. They may occur as an isolated, single pulse (e.g., an impulse acting on a rollbar (Figure 10.3.1), or as a train of separate or overlapping pulses. The pulses forming the train may be periodically repetitive (e.g., worn machine bearings, Figure 10.3.2), controlled (e.g., laser cleaning of materials) or random (e.g., railtrack faults from a moving train, Figure 10.3.3). Overlapping pulses may involve only two short-duration, partially overlapping pulses (e.g., circuit breaker drive mechanisms, Figure 10.3.4) or may involve many prolonged duration pulses producing quasi-steady, complex periodic signals (e.g., rotating machine drive interface, Figure 10.3.5).

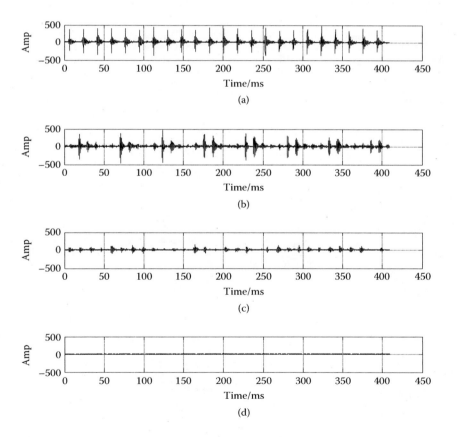

FIGURE 10.3.2
Vibration signals produced by machine bearings with different faults: (a) outer ring, (b) inner ring, (c) ball, (d) none.

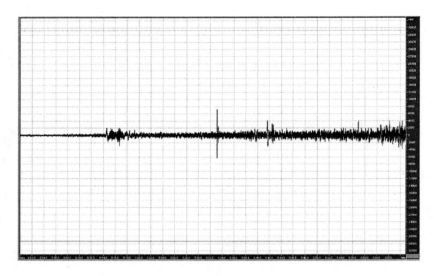

FIGURE 10.3.3
Time-varying acoustic signals produced from rail tracks by a moving train. (From Deakin, A. G., Rallis, I., Zhang, J., Spencer, J. W., and Jones, G. R. (2005). Towards holistic chromatic intelligent monitoring of complex systems in a chromatic approach to complexity. *Proceedings of the Complex Systems Monitoring Session of the International Complexity*, Science and Society Conference. Liverpool. With permission.)

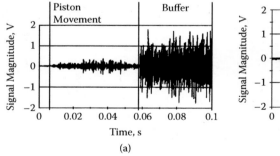
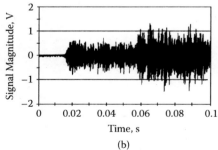

FIGURE 10.3.4
Acoustical signals from a high voltage circuit breaker interupting (a) no current, (b) 31.1 kA fault current. ((a) From Jones, G. R. (2001). Optical fiber based monitoring of high voltage power equipment in *High voltage engineering and testing*, Ryan, H. M., Ed., 2nd ed., IET., chap. 21. With permission, (b) from Russell, P. C., Cosgrave, J., Tomtsis, D., Vourdas, A., Stergioulas, L., and Jones, G. R. (1998) Extraction of information from acoustic vibration signals using Gabor transform type devices. *Meas. Sci. Technol.*, 9, pp. 1282–1290. With permission.)

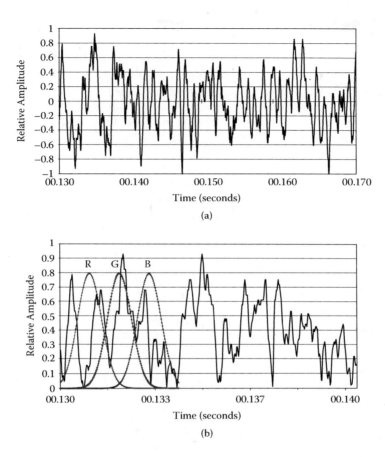

FIGURE 10.3.5
Example of periodic signals: (a) raw acoustic signal, (b) rectified signal with chromatic filters
R, G, B overlaid.

The chromatic approach may be adapted to address each of these types of signals in either the frequency or time domains and guided by the flowchart of Figure 10.2.1.

In the case of a train of periodically repetitive pulses, the time period covered by three time-domain processors is advantageously arranged to correspond to the cycle time of the combination of pulses (Figure 10.3.2). In the frequency domain the signal sampled should also have a time window covering the cycle time.

For the case of a single isolated pulse (Figure 10.3.1), the chromatic processors may be deployed to cover the entire pulse duration to give an integrated pulse signature. In the time domain this requires the three chromatic processors to have width that, together, cover the total pulse duration. In the frequency domain the sampling of the signal should have a time window covering the pulse duration and being compatible with Nyquist requirements (e.g., Morinaga et al., 2000).

If a more detailed assessment of the pulse structure is needed, frequency-domain processors would need to sample a series of time windows during the pulse. Time-domain processors would need to be time stepped along the pulse. The size of time stepping or frequency windowing would depend upon the details sought and the complexity of the pulse structure (e.g., two marginally overlapped pulses (cf. Chapter 6, Section 6.2.3.2) or several long-duration pulses forming a steady periodic signal).

Randomly occurring pulses need first to be detected by thresholding in the time domain (Figure 10.2.1) followed by a suitable choice of procedures to deal with the possible complex features of the pulse. These may involve using combinations of time- and frequency-domain processing within up to three different frequency ranges.

Particular examples of these various deployments of chromatic processing are given in the following sections.

10.4   Examples of Chromatic Processing Deployments

10.4.1   Simulated Pulse Trains (Frequency Domain)

An example of the use of frequency-domain chromatic processing with software triangular filters and Fourier Transformation (Figures 10.2.1 (iii) and 10.2.2 (vi)) for the retrospective processing of periodic pulsatile signals is in tests to simulate the faulty operation of machine bearings. Such bearings consist of ball bearings that run along a path between a fixed outer ring and a rotating inner ring. The development of faults on the ball bearings and their enclosure can produce major failures of a machine. The movement of the ball bearings during machine operation produces vibration signals from which it may be possible to detect the inception of a fault regime.

Figure 10.3.2 shows a group of vibration signals measured on a bearing case cover during the operation of a machine in which various defects were simulated in sequence on the outer ring (Figure 10.3.2a), the inner ring (Figure 10.3.2b), and the ball bearing itself (Figure 10.3.2c). Figure 10.3.2d shows a record from a normal ball-bearing structure.

The vibration signals shown in Figure 10.3.2a–d are Fourier transformed to yield the four frequency-domain signals shown in Figure 10.4.1a. The frequency-domain signal covering the range 0–5 kHz is addressed with three nonorthogonal triangular R,G,B processors (Figure 10.4.1a) to yield the chromatic coordinates H, L, S. Each signal (a)–(d) of Figure 10.3.2 is processed in the same manner, and the points representing their coordinates for the normal operating condition and those of the artificially induced faults are shown collectively on the H-L, H-S polar diagrams of Figure 10.4.1b, c. The results of several tests under each condition are shown.

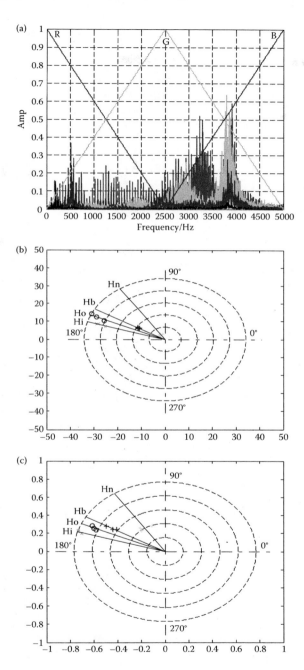

FIGURE 10.4.1
(See color insert following page 18). Frequency domain processing of acoustical signals from machine bearings. (a) Fourier-transformed signals showing *R, G, B* triangular filter superimposed. (b) *H-L* polar diagram, (c) *H-S* polar diagram (Faults: o—outer ring (Ho), x—inner ring (Hi), +—ball (Hb), *—normal (Hn)).

All fault conditions are distinguishable from normal operation (L ≅ 0.1) in having higher values of L (signal strength), the most pronounced effects being those of faults on the outer and inner rings (0.8 < L < 1.0) (Figure 10.4.1b). The outer ring fault is distinguishable from the other three conditions in having a lower S value (signal spread) (0.4 < S < 0.6) compared with (0.8 < S < 1.0) (Figure 10.4.1c). The inner ring fault is distinguishable in having a higher H (dominant frequency) than the other conditions (i.e., HI > HO, HB > HN, Figure 10.4.1c). Consequently, a faulty condition may be chromatically distinguished from the normal condition, and rules based upon the relative value of the chromatic parameters H, L, S can be assembled for distinguishing between the three faults considered.

10.4.2 Industrial Plant Faults (Frequency Domain)

An illustration of the use of frequency-domain chromatic processing with hardware filters (Section 10.2.1 (i)–(ii)) for processing pulsatile signals is for monitoring a Combined Heat and Power (CHP) water pump online continuously for a period of several months (Russell et al. 2000; Deakin et. al., 2005; Cooke, 2000). The vibration signals were detected with an optical fiber homodyne interferometer (Cooke, 2000).

The system produces a continuous stream of H, L, S coordinate values, which are mapped on H-L, H-S polar diagrams (Figure 10.4.2). The H-L diagram shows a migration of the operating area away from L = 0 at the beginning of the monitoring period toward L ~ 0.8 over the duration of the monitoring period, at which point the pump failed. The H-S diagram shows

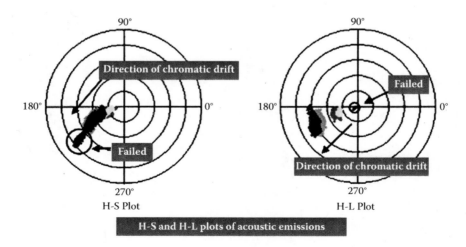

FIGURE 10.4.2
(See color insert following page 18). *H-S, H-L* polar diagrams for acoustic signals from a water pump at a combined heat and power plant recorded over a period of 5 months (greyscale distinguishes calendar month). (From Cooke, R. (2000). An intelligent monitoring system, Ph.D. Thesis, University of Liverpool).

a similar migration before reverting to S = 0 after the failure occurred. There is also evidence of a drift in the average H value to somewhat higher levels. The high H, L, and S values following fault development are consistent with the result of the simulated tests with the bearings. The difference in S values between the normal simulation test and the initial conditions for the CHP tests is probably associated with background noise interference at low signal strengths particularly in the robust CHP environment.

The implication of Figures 10.4.1 and 10.4.2 is that, although the sensing and chromatic processing are different in the two sets of tests, the processed data can identify bearing-based machine faults under ideal operating condition and the evolution of faults under robust industrial conditions. Once the faults under the industrial conditions become well developed, the chromatic signatures appear similar to those in the simulation tests.

10.4.3 Diesel Engine (Time Domain)

An example of the deployment of fixed time-domain chromatic processing with Gaussian nonorthogonal filters for addressing repetitive pulsatile vibration signals is in monitoring the operation of a four-cylinder diesel engine.

Analysis of vibration signals from the surface of the cover of a diesel engine is a means for monitoring the condition of the engine despite the complexity of its structure and operating mechanisms (Zhang et al., 2003). Typical time-based vibration signals from the cover of such a diesel engine close to its fourth cylinder are of a similar form but with different details to the periodic pulse trains produced by the machine bearings (Figure 10.3.2). Acoustical signals were obtained under normal operation of the engine and for a number of simulated fault conditions on the fourth cylinder (cylinder and piston wear (blow by), injection clogging, needle valve, and cylinder nozzle wear).

Figure 10.4.3a shows typical pulsatile signals that are emitted during the 70-ms cycle duration of the engine with three fixed time domain processors (R, G, B) superimposed. The points corresponding to the H, L coordinates for several signals and for each of the simulated conditions are shown collectively on the H-L polar diagram of Figure 10.4.3b. Each of the four groups are clearly distinguishable from their H-L coordinates. Normal operation (H_n) corresponds to L ~ 0; high injection pressure to H_h ~180°, 0.2 < L < 0.4; low injection to H_l ~340°, 0.2 < L < 0.3; blow by to H_b ~210°, 0.6 < L < 1.

The pulsatile signals for the tests performed on the diesel engine can also be considered in terms of principal component analysis (PCA) (Jollife, 1986) using the average energy in the sequence of pulses. The result of such a PCA shows that the test results fall into four groups corresponding to each operational condition (Figure 10.4.4) so distinguishing between normal operation and each of the faults as does the fixed time-domain chromatic processing (Figure 10.4.3b). However the chromatic approach provides a better level of traceability in indicating which features of the signals are responsible for the differentiation (e.g., L—energy content of the pulse train, H—dominant time during the pulse cycle period, S—the spread of pulses throughout the cycle period).

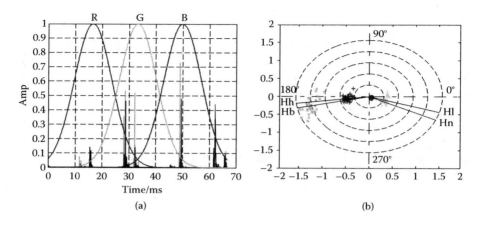

(a)   (b)

FIGURE 10.4.3

Fixed time domain processing of acoustical signals from a diesel engine: (a) one cycle of pulses with fixed time *R, G, B* Gaussian processors superimposed, (b) *H-L* polar diagram. (o—normal, x—blow by, +—high injection pressure, *—low injection pressure) (Engine data courtesy of Dr. B. Liu, National University of Singapore. With permission.)

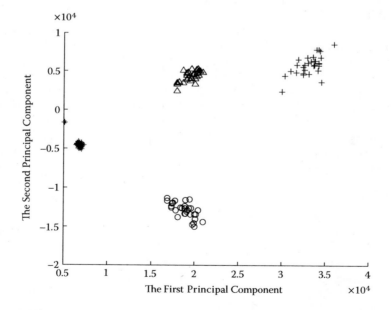

FIGURE 10.4.4

(See color insert following page 18). Principal components analysis of diesel engine faults from acoustic signal. (*—normal condition, +—blow by, Δ—high injection pressure, o—low injection pressure) (From Zhang, X., Wen, G., and Li, X. (2003). Vibration feature abstraction and classification of diesel faults. *Int. J. Plant Eng. Manage.* Vol. 8, No. 1. With permission.)

10.4.4   Controlled-Pulse Train (Frequency Domain)

The use of frequency-domain processing with a single time window that covers the entire pulse duration is illustrated with the acoustical signals that are produced during the processing of a material's surface by a high-power laser pulse (e.g., Lee et al., 1999). In this case the chromatic processing is with x-y algorithms (Chapter 3, Section 3.3).

A series of individual, short-duration laser pulses (each 5–20 ms) are used to remove surface film from contaminated surfaces. During interaction of the laser pulse with the surface, a plasma plume is formed that expands rapidly, creating shock waves that are detectable acoustically (e.g., Lu and Aoyagi, 1995). The resulting acoustical signal has the potential for tracking the progression of the processing (e.g., Lee et al., 1997) to indicate the completion of the processing. However, the use of techniques based upon acoustic intensity alone are susceptible to interference from spurious effects. Because the x, y chromatic parameters are intensity independent, the chromatic approach offers a means for overcoming the difficulties with intensity-only methods (Lee et al., 1999).

Each acoustical pulse (1 ms duration) produced by each of nine separate laser pulses in cleaning photocopier toner paper (Lee et al., 1997) is addressed by chromatic processors covering the acoustic frequency range 2–12 kHz and the time duration of the pulse. This produces a series of x, y parameter values as shown in Figure 10.4.5a. The series of points (1–9) corresponding

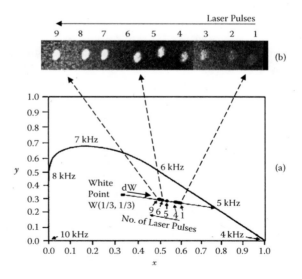

FIGURE 10.4.5
Comparison of chromatic acoustical signatures with extent of laser treatment of toner paper: (a) variation of acoustical signals on an x-y chromatic diagram, (b) Sequence of images showing incremental removal of toner by a series of laser pulses. (From Lee, J. M., Watkins, K. G., Steen, W. M., Russell, P. C., and Jones, G. R. (1999) Chromatic modulation based acoustic analysis technique for in-process monitoring of laser materials processing. *J. Laser Appl.*, Vol. 11, No. 5, pp. 199–205. With permission.)

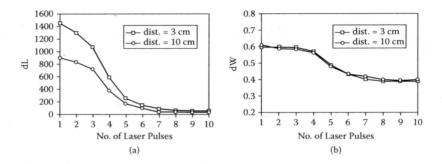

FIGURE 10.4.6
Dependence of acoustical chromatic parameters upon the number of laser pulses and location of transducer relative to the processed surface: (a) chromatic L (effective signal intensity), (b) Chromatic dW (signal purity). (From Lee, J. M., Watkins, K. G., Steen, W. M., Russell, P. C., and Jones, G. R. (1999). Chromatic modulation based acoustic analysis technique for in-process monitoring of laser materials processing. *J. Laser Appl.*, Vol. 11, No. 5, pp. 199–205. With permission.)

to each of the nine independent laser pulses are aligned between the achromatic point (1/3, 1/3) and the 5-kHz monochromatic point, the location of points being displaced towards the achromatic point as the laser cleaning progresses (Figure 10.4.5a). This corresponds to a substantial reduction in the purity of the acoustical signal dW ~ 0.6–0.4 (Figure 10.4.5a) so that the numerical value of the purity parameter W provides a metric for the progress of cleaning (Figure 10.4.5b).

The insensitivity of the purity parameter W to acoustic intensity variations is illustrated by the coincidence of W data gathered from two microphones, each at a different distance (3, 10 cm) from the surface being cleaned (Figure 10.4.6b). Conversely, the L values (corresponding to the energy in the acoustic signal) depend upon the location of the microphones (Figure 10.4.6a).

10.4.5   Double Pulse Signals (Frequency Domain)

The use of gated H, L, S chromatic processing in the frequency domain for acquiring information contained within a section of a double pulse is illustrated for a vibration signal produced during the operation of a high-voltage circuit breaker (Russell et al., 1998). High-voltage circuit breakers (HVCB) are physically large devices for switching high electric currents (tens of kiloamperes) at high voltages (hundreds of kilovolts) on electric power transmission networks (Jones, 2001) and form the ultimate protection for such networks.

They operate by mechanically separating two metallic contacts across which the high voltage and electric current produce an electric plasma (Chapter 4, Section 4.3.2), which is then quenched by a supersonic flow of gas produced by a mechanical piston so that the current is interrupted. Mechanical vibrations are produced first by the piston action, gas flow, etc., and also when the piston strikes a buffer at the end of its stroke. Acoustical signals extending over tens of milliseconds are produced that contain information about the condition and operation of the circuit breaker.

Vibration signals obtained with an optical fiber interferometer from such a circuit breaker are shown in Figure 10.3.4a, b during operation with no fault current and with a moderately high fault current of 31.1 kA (Jones, 2001, Russell et al., 1998). The central time portion of the signal (~ 50 ms) corresponds to the movement of a mechanical piston for producing the arc quenching gas flow, whereas the later portion corresponds to the buffer arresting the piston. Comparison of the signals at 0 kA and 31.1 kA conditions during the piston movement period shows qualitatively a significant difference that can be quantified by applying frequency-domain processors to a number of gated sections of the piston movement period.

The time variation of the H, L chromatic parameters derived for each time window during the piston movement period (C, time windows 6–12) are shown in Figure 10.4.7a, b. The chromatic S parameter had values in the range

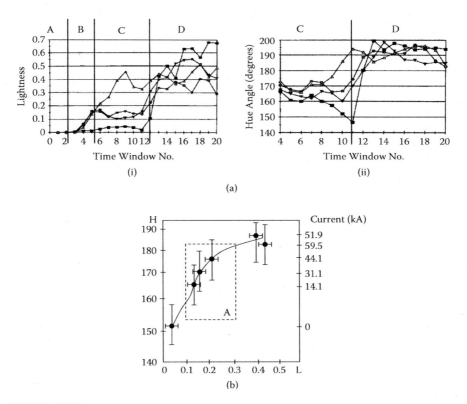

(a)

(b)

FIGURE 10.4.7
Frequency-domain chromatic parameters. (a) Time variation of chromatic parameters for various fault currents (i) *L*, (ii) *H* [■ 0 kA ▼ 14.3 kA ● 31.1 kA ▲ 59.5 kA] (b) *H-L* chromatic diagram (piston operation period C). (From Russell, P. C., Cosgrave, J., Tomtsis, D., Vourdas, A., Stergioulas, L., and Jones, G. R. (1998). Extraction of information from acoustic vibration signals using Gabor transform type devices. *Meas. Sci. Technol.*, 9, pp. 1282–1290. With permission.)

0.92–0.98 for each operating condition, indicating the signals to be highly monochromatic. Variations in the H, L parameter values are different during the piston movement phase for the current ranges shown.

Average values and spread for H and L corresponding to different peak currents are compared on the Cartesian H-L diagram of Figure 10.4.7c. This shows a trend for both H and L to increase with fault current for the fault-current regime lying within region A of Figure 10.4.7b. Consequently, it is possible to track the mechanical severity duties to which the circuit breaker has been exposed (particularly those up to a range limit of 50–60 kA) so that servicing schedules can be planned.

10.4.6   Chromatic Infrastructure of a Single Pulse (Frequency Domain)

The use of stepped frequency domain, H, L, S chromatic processing (Figure 10.2.1) for acquiring detailed information embedded in an acoustic pulse is illustrated with an isolated acoustic pulse produced by a mechanical impulse applied to an industrial rollbar. The motivation for such monitoring is for identifying the early onset of mechanical failure.

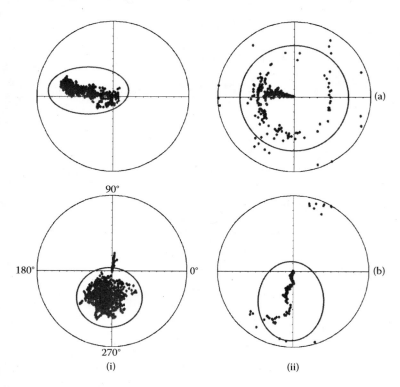

FIGURE 10.4.8
(See color insert following page 18). Chromatic analysis results for the industrial rollbars: (a) a good-quality rollbar (b) a bad-quality rollbar. (i) *H-S*, (ii) *H-L*.

Time-stepping (~1ms) frequency domain R, G, B chromatic filters of width 10ms along the pulsatile signal from a faultless rollbar shown in Figure 10.3.1a produces trajectories of points on H-S and H-L polar diagrams (Figure 10.4.8a (i), (ii)). The series of points form distributed patterns with discernible identities. On the H-S diagram, the points cluster along a radius inclined at H ~ 160°, whereas on the H-L diagram many points are scattered along a circle of radius L ~ 0.45.

Applying the same chromatic processing to a faulty rollbar produces different patterns of points distribution on H-S and H-L polar diagrams (Figure 10.4.8b (i), (ii)). In this case the points on the H-S polar diagram cluster in a sector defined by 230° < H < 300°. On the H-L diagram they cluster mainly along a radius inclined at about H ~ 260°.

Thus, the differences between the H-S, H-L distribution of points is sufficiently clear for distinguishing between a good and a faulty rollbar.

10.4.7 Narrow-Frequency-Band Continuous Signals (Frequency Domain Plus Second Generation)

An example of time-stepping frequency domain processors to produce H, L, S chromatic parameters, which are then addressed by second-generation chromatic processing (Figure 10.2.1), is illustrated with complex continuous periodic signals from an industrial machine-drive interface rotating at high speed. The requirement is to identify fault development in such machine-drive interfaces.

The complex structure of acoustical signals from such machine-drive interfaces sampled at 48 kHz is shown in Figure 10.3.5a. The signal is rectified before being filtered to cover a limited frequency band (e.g., 500–1,500 Hz) and R, G, B frequency domain chromatic processors time stepped along the signal (Figure 10.3.5b). The R, G, B outputs are chromatically processed (Equations 1.2–1.7, Chapter 1) to produce H-S, H-L polar diagrams.

The pattern of points on H-S, H-L diagrams produced by such time stepping form complex distribution patterns that may contain features that could distinguish a healthy machine-drive interface (Figure 10.4.9a) from a damaged one (Figure 10.4.9b).

Often, the distinguishing features may not be immediately apparent but may be enhanced by secondary chromatic processing. For example, three chromatic processors can be applied to the pattern of points represented on an H-L Cartesian diagram to produce secondary H_S, L_S, S_S, which quantify the difference between the healthy and damaged machine-drive interfaces as shown in Table 10.1. The differences between the H and S values for the good and faulty situations (30°/10°, 0.7/0.55) are sufficient for discriminating between the conditions.

10.4.8 Random Pulsatile Events (Combined Frequency and Time Domain)

Addressing random pulsatile events requires the deployment of more of the chromatic processing options given in Figure 10.2.1 in combination with

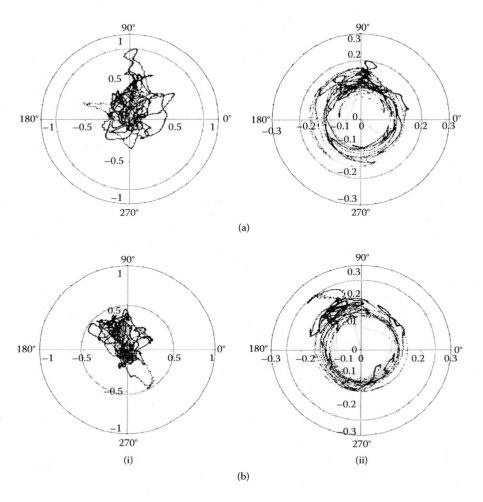

FIGURE 10.4.9
H-S and *H-L* diagrams for good and bad machine-drive interfaces, zooming in to a 1 kHz bandwidth: (a) healthy machine-drive interface, (b) damaged machine-drive interface (i) *H-S*, (ii) *H-L*.

TABLE 10.1

Comparison of Second-Generation *H, L, S*

Machine-Drive Interface (Figure 10.4.9)	HS	LS	SS
Good (a)	30°	0.09	0.70
Faulty (b)	10°	0.11	0.55

each other. There is first a need to detect relevant pulses by appropriate time-domain thresholding, followed by both frequency- and time-domain processing. The frequency-domain processing is undertaken in three different frequency bands, and the results of these analyses are combined in a secondary chromatic processing to estimate the severity and type of fault that produced the acoustic pulse.

An example of the deployment of the chromatic approach for addressing randomly occurring pulses is for detection of rail-track faults from a traveling railway engine. The movement of the engine along the rail track produces vibration of the metallic rails via the engine wheels, which in turn produce

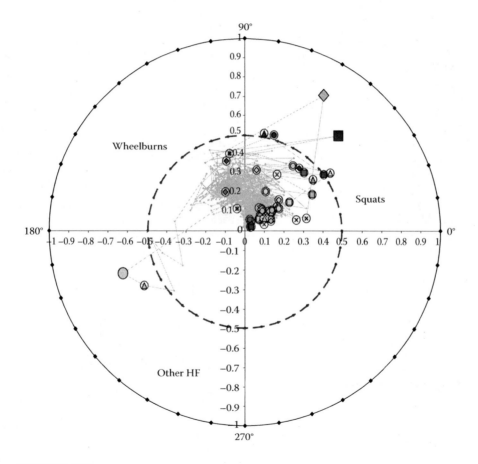

FIGURE 10.4.10
H-L polar diagram from two-stage chromatic processing of railtrack signals (inner circle—threshold for event detection). (From Deakin, A. G., Rallis, I., Zhang, J., Spencer, J. W., and Jones, G. R. (2005). Towards holistic chromatic intelligent monitoring of complex systems in a chromatic approach to complexity. *Proceedings of the Complex Systems Monitoring Session of the International Complexity*, Science and Society Conference. Liverpool. With permission.)

acoustic signals detected by microphones mounted on the engine. Embedded in these signals are components produced by defects on the rail that need to be discriminated from the various other components of the signal.

An example of a random pulsatile signal obtained during rail-track monitoring has been given in Figure 10.3.3 for a train accelerating from a station. Two-stage chromatic processing of the acoustic signals yields trajectories on an H-L polar diagram of the form shown in Figure 10.4.10. Three of the four quadrants of the H-L polar diagram correspond to different types of faults. Excursions by loci of analyzed signals into these quadrants above an L threshold level are indicative of the development of a particular type of fault.

10.5 Summary

Various chromatic-processing procedures have been described that emphasize the choice between time and frequency domain processing, the use of H, L, S and x, y chromatic diagrams, time stepping in the time domain, second-generation chromatic processing, etc. The use of hardware and software processing has been compared.

Different forms of acoustic pulsatile signals have been identified. These range from trains of pulses that have a periodically cyclical form, via isolated pulses, to randomly occurring pulses.

Deployments of the various chromatic processing methods for addressing the different pulsatile situations are described. These range from processors addressing the cyclical period of several pulses to individual pulses. Finer-scale processing of the structure of a pulse is considered, ranging from pulses containing two partially overlapping pulses to long-duration continuous signals. The manner in which secondary processing can be used to distinguish detailed differences in pulses from normal and faulty conditions of the source is indicated.

A number of different situations that produce acoustic pulses of various kinds are described, including machine bearings, diesel engines, circuit-breaker operation, mechanical rollbar testing, and rail-track faults.

References

Cooke, R. (2000). An Intelligent Monitoring System. Ph.D. thesis, University of Liverpool.

Deakin, A. G., Rallis, I., Zhang, J., Spencer, J. W., and Jones, G. R. (2005). Towards holistic chromatic intelligent monitoring of complex systems in a chromatic approach to complexity. *Proceedings of the Complex Systems Monitoring Session of the International Complexity*, Science and Society Conference. Liverpool.

Jollife, I. (1986). *Principal Component Analysis*. Springer-Verlag, New York.

Jones, G. R. (2001). Optical fibre based monitoring of high voltage power equipment, in *High Voltage Engineering and Testing*, Ryan, H. M., Ed., 2nd ed., IET, chap. 21.

Jones, G. R., Russell, P. C., Vourdas, A., Cosgrave, J., Stergioulas, L., and Haber, R (2000). The Gabor transform basis of chromatic monitoring. *Meas. Sci. Technol.*, 11, 489–498, 2000.

Lee, J. M., Watkins, K. G., and Steen, W. M. (1997). Investigation of acoustic monitoring in the laser cleaning of copper. *Proc. ICALEO '97* (Laser Institute of America, Orlando, FL), Section C, pp. 226–233.

Lee, J. M., Watkins, K. G., Steen, W. M., Russell, P. C., and Jones, G. R. (1999). Chromatic modulation based acoustic analysis technique for in-process monitoring of laser materials processing. *J. Laser Appl.*, Vol. 11, No. 5, pp. 199–205.

Lu, Y. F. and Aoyagi, Y. (1995). Acoustic emission in laser surface cleaning for real time monitoring. *Jpn J. Appl. Phys.*, 34, L1557–L1560, Part 2, No. 11 B.

Morinaga, N., Kohno, R., and Sampei, S. (2000). *Wireless communications technologies: New Multimedia Systems*. Kluwer Academic.

Russell, P. C., Cosgrave, J., Tomtsis, D., Vourdas, A., Stergioulas, L., and Jones, G. R. (1998). Extraction of information from acoustic vibration signals using Gabor transform type devices. *Meas. Sci. Technol.*, 9, pp. 1282–1290.

Russell, P. C., Deakin, A., Jones, G. R., Sproston, J. L., White, M. D., and Owen, I. (2000). Classification of complex time varying signals using chromatic compression and Kohonen networks, Intelligent Systems and Applications (ISA 2000), Wollongong, Australia, December 11–15.

Zhang, X., Wen, G., and Li, X. (2003). Vibration feature abstraction and classification of diesel faults. *Int. J. Plant Eng. Manage.* Vol. 8, No. 1.

11

Chromatic Monitoring of Activity and Behavior

A. Koh, K.J. Wong, and S. Xu

CONTENTS

11.1   Introduction

There are situations that demand the addressing of the activity and behavior of an individual to be undertaken for the benefit of the individual but unobtrusively and without invasion of privacy. Such situations make different demands upon sensing and monitoring not only because of the need for no intrusion but also because often there are physiological aspects involved, adding to the complexity of the problem. Furthermore, there are significant cost constraints to be observed.

Examples of such activity and behavior situations are

- The level of activity of the elderly and infirm to guard against the consequences of health damage falls in domestic environments (Xu and Jones 2006, Xu 2004, Wong 2006).
- The level and increase in fatigue of a road vehicle driver to avoid causing road accidents (Koh et al. 2007).

In the former case, both spectral and space-domain chromaticity, coupled together, are used (Xu and Jones 2006, Xu 2004, Wong 2006). Also spatial chromaticity from both optical and infrared instruments are combined (Wong 2006). Such an approach not only provides an indication of the general mobility of an individual and the severity of a fall but also indications of room temperature and the other behavior-affecting artifacts (e.g., television on/off, curtains opened/closed, etc.). The unobtrusivity and privacy requirements are satisfied by it not being necessary for the individual to carry or activate a tag, and that it is only specific points in the living environment, and not an individual, that are being addressed. Changes in the environment due to inanimate objects, light, etc., are distinguishable from changes caused by an individual.

In the case of a fatigued vehicle driver, both physiological aspects (driver fatigue pattern and response) and vehicle reaction need to be taken into account (Koh et al. 2007). The driver response and vehicle reaction can be monitored from signals from a gyroscope mounted on the vehicle dashboard to monitor the lateral movements of the vehicle. Signals peculiar to a driver emerging from microsleep may be distinguished from other signals via time-stepped chromatic analysis. Physiological sleep patterns may also be described in terms of chromatic parameters. Such effects may be taken along with the gyroscope signals to provide a chromatically based probability-of-fatigue score.

This chapter describes the chromatic techniques used to address such activity and behavioral aspects.

11.2 Activity and Behavior of a Vulnerable Individual

11.2.1 Applications of Optical Wavelength and Space Chromaticity

11.2.1.1 Introduction

There are many situations in which the position or movement of objects need to be monitored unobtrusively. An approach is described whereby an enclosed environment is categorized via the optical chromatic signature of a limited number of points fixed within that space. The movement of an object within the space may be tracked from the sequential changes in the chromaticity of these individual points. Subsequent analysis is undertaken by ascribing zones to the various detection points and using space-based

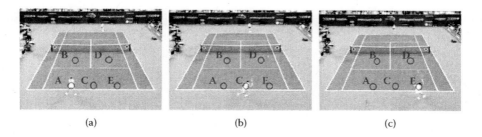

(a)   (b)   (c)

FIGURE 11.2.1.1
(See color insert following page 18). Chromatic position monitoring frames (a), (b), (c) show how a player changes the chromatic signature of monitoring points A, C, and E, respectively, when moving across the tennis court. (From Wong K., Xu S., and Jones G. R. (2005). Chromatic identification of complex movement patterns, *Proc. Int. Conf. on Complexity in Sci., Med. Sociol.,* Centre for Complexity Research, Liverpool. With permission.)

chromaticity for mapping the movement pattern. Figure 11.2.1.1a, b, and c show images of part of a tennis court with a player who intermittently moves and changes position. A limited number of points on the court are identified as significant in identifying the movement pattern of the player, and the chromatic signatures of these points are individually noted. Movements of the player across these points cause the chromatic signature of each point to sequentially change (e.g., from blue to white). By tracking the position of these signature changes, the movement of the player shown in Figure 11.2.1.1a, b, c is obtained.

11.2.1.2 Description of Monitoring System

Within a closed environment, a limited number of critical locations can be identified, each with a particular chromatic signature (Figure 11.2.1.2a). These locations may be points on a door, chair, television, window, floor area, etc. The points are addressed by a colocated and remote array of optoelectronic sensors with a chromatic discrimination capability and with each sensor associated with one of the critical locations. The outputs from the sensors are forwarded to a central control unit for processing. Each sensor produces three output signals (R, G, B) from which the chromaticity of the point addressed is determined in HSL chromatic space using equations Chapter 1, Section 1.2–1.7).

A change in the chromaticity of a point indicates a physical change. The nature of which depends upon the functionality of that location. For instance, a door-located point indicates if the door is opened or closed; a television point indicates if the television is on or off; a point high on a wall responds to different light conditions; a chair point indicates chair occupancy. Sequential eclipsing of a series of points on the floor is indicative of a movement across the floor from which patterns of movement and behavior can be derived [Xu and Jones 2006, Xu 2004]. The process may be compared to the points located by a Radar-tracking system (Figure 11.2.1.2b), except that the points are detected optically.

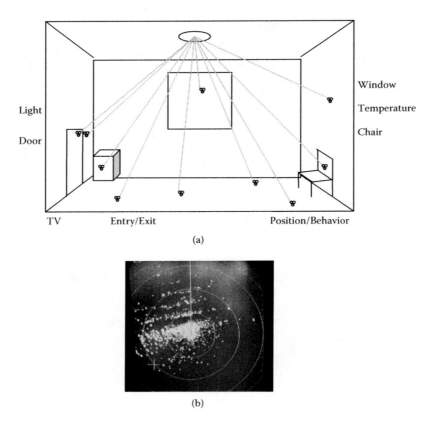

(a)

(b)

FIGURE 11.2.1.2
Basic principle of the position-addressing optical chromatic system: (a) system layout showing a limited number of chromatically identified points, (b) examples of RADAR system image.

Furthermore, it should be appreciated that it is only a limited number of selected points in the environment that are monitored and not an individual himself/herself. Movement and behavior are determined from an individual's effect on these points, which can be selected so as not to intrude upon the privacy of the individual.

An essential part of the monitoring is to identify a change in the status of a location by a sufficient change in the total chromaticity of the location above a predetermined level. It is not essential to demand that the value of the changed chromatic state be determined but simply that it has exceeded unambiguously the predetermined threshold. However, once the perturbation has passed, it is essential to know that the chromatic status of the point has returned to its original, predetermined value. In the event of that not occurring, then there has been a change in the background environment, which needs to be accommodated in conjunction with any or all of the other points in the environment that may be simultaneously affected. For example, global light fluctuations would be identifiable by groups of non-afinely related points being affected simultaneously.

The determination of movement/behavioral patterns and the discrimination of effects not associated with an individual (e.g., ambient light changes, etc.) may be facilitated by further chromatic processing of the optical chromaticity changes in the chromatic space domain [Xu and Jones 2006, Xu 2004] as described in Section 11.2.3.1.3 (iii)–(iv).

11.2.1.3 Chromatic Changes at Selected Locations

(i) Total Chromatic Change

A perturbation of a chromatically addressed point in the environment of Figure 11.2.1.2a generally causes a change in one or more optical chromatic parameters H, L, and S. It is therefore appropriate, for identifying whether a location has suffered a chromatic change, to consider whether there has been a change in the chromaticity vector defined by Equations 1.8–1.12 in Chapter 1, Section 1.3.2(a) in terms of the three parameters H, L, and S together.

A meaningful event at a selected location is deemed to have occurred if the total chromatic change as defined by Equations 1.8–1.12 exceeds a specified threshold.

(ii) Spatial Chromaticity

The locations selected to be addressed (Figure 11.2.1.2a) can be categorized in various ways based on the information being sought. As a simple example, a number of points located on the floor of a room are shown in Figure 11.2.1.3 (a) may be used for addressing movement. To indicate the relative distances between various locations, the selected locations are sorted in an ascending sequence according to their distances from the top-left corner of the planar monitored area. The location indicators (1,2,3, etc.) form elements of a discrete data set as described in Chapter 2 Section 5, which can be addressed by three nonorthogonal processors R, G, and B (Figure 11.2.1.3b). The widths of the R and B filters are determined by the number of selected locations N and are half of the G filter. If the number of selected $(N-1)/2$ locations N is more than 3, the width of the R and B filters are set to $N/2$ for N even and $(N-1)/2$ for N odd. For $N \le 3$, the R and B filter widths are 1.

The output (x_0) from each of the chromatic processors (x_p) is of the form (Appendix 1.I, Chapter 1)

$$x_0 = \sum_{n=0}^{N} A_n X_p \tag{1}$$

where A_n is the amplitude of the signal at location n, and N is the total number of locations. The precise nature of A_n depends upon the information being sought. In the simplest case it may be binary (0—no visit, 1—visits) to indicate whether a site is being visited at any time instant.

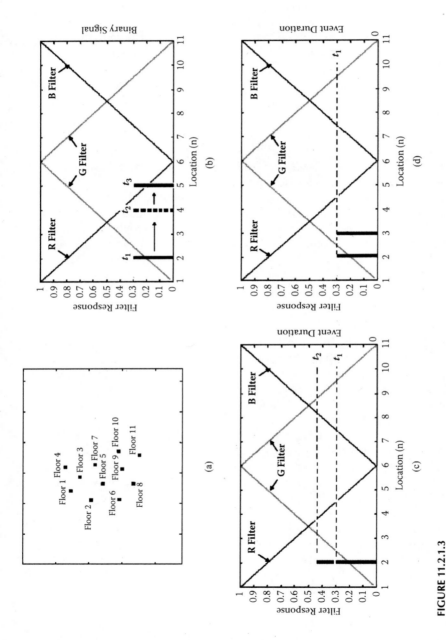

FIGURE 11.2.1.3

Selected locations (1–11) and event occurrence: (a) location of selected points, (b) ordered locations (n) versus filter response and time shift of a binary event signal, (c) ordered locations (n) versus filter responses and event duration clock, (d) ordered locations (n) versus filter responses and event spanning two location simultaneously. (From Xu S. and Jones G. R. (2006). Event and movement monitoring using chromatic methodologies, *IoP J. Meas. Sci. Tech.*, 17, 3204–3211. With permission.)

Transforming R_o, G_o, B_o into spatial H, L, S leads to H indicating the dominant visit location, L the intensity of activity, and S the distribution of the visits.

(iii) Chromatic Representation of General and Localized Movement

The responsivities of the processors R, G, B (Figure 11.2.1.3b) with respect to their maximum values are arranged to be identical so that $R_{MAX} = G_{MAX} = B_{MAX} = 1$. The responsivity of each pair of processors in their overlap region is linear and complementary so that

$$R_n + G_n = 1, \quad B_n = 0 \quad 0 \leq n \leq N/2 \tag{2}$$

$$G_n + B_n = 1, \quad R_n = 0 \quad N/2 \leq n \leq N \tag{3}$$

Hence, for a purely monochromatic signal of unit amplitude at a location n, equations 1.2–1.7 (Chapter 1, Section 1.3.1) show that S = 1, H = H_n, L = 1/3, i.e., S, L are independent of n, whereas H indicates the location of an event (n) in the environment as the location varies with time (Figure 11.2.1.3b).

To quantify any immobility periods (i.e., time durations of continuous residence at one location), each location may be allocated a clock to record the time duration of an event at that location. The clock commences to count once an event is identified at that location and continues to count as long as the event remains detectable without interruption at that location. When the event ceases to be detectable at that location, the clock ceases to count and is cleared. Consequently, the clock is an "event timer." The amplitude A_n (Equation 1) is then made to correspond to the event duration so that an "event duration: position" spectrum is formed (Figure 11.2.1.3c). The spectrum is then addressed by the three nonorthogonal processors R, G, B to yield the spatial chromatic parameters H_s (the dominant location of the event), S_s (the spread of locations at which the event occurred), and L_s (the effective event duration).

For the case shown on Figure 11.2.1.3c, an event occurs confined to a single location (n = 2), which corresponds to a highly monochromatic spectrum (S_s = 1). Also from Equations 1, 2, 3, and Equation 1.3 in Chapter 1, Section 1.3.1.1,

$$L_s = (R_o + G_o + B_o)/3$$

$$= t_n \cdot (R_n + G_n + B_n)/3$$

$$= t_n/3 \tag{4}$$

i.e., L_s is proportional to the time duration of the event at location n.

In practice, it is possible for an event to affect two adjacent locations simultaneously if the locations are sufficiently close, e.g., Figure 11.2.1.3d with an event covering both locations 2 and 3 simultaneously. In such cases, the L_s value needs to be normalized with respect to the number of locations involved. For example, in the case of Figure 11.2.1.3d, following the procedures of Equations 2, 3, and 4, L_s = 2t/3, which when normalized for two locations,

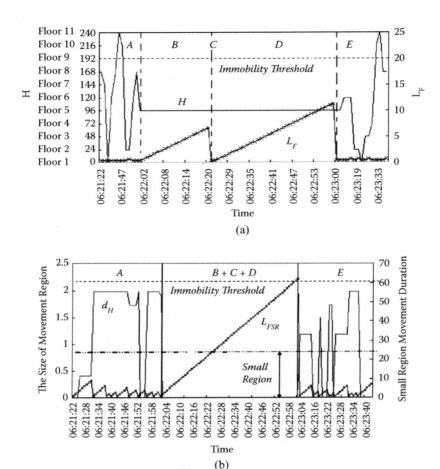

FIGURE 11.2.1.4
(See color insert following page 18). Time variation of parameters derived from chromatic processing: (a) Hs and L_F versus time—general movement, (b) movement size (d_H) and small-region-accommodated immobility (L_{FSR}) versus time. (From Xu S. and Jones G. R. (2006). Event and movement monitoring using chromatic methodologies, *IoP J. Meas. Sci. Tech.*, 17, 3204–3211. With Permission)

gives the normalized $L_F = L_s/2 = t/3$. L_F then retains a strict proportionality to time duration, and an L_F : time graph can be produced as shown in Figure 11.2.1.4a. This indicates how long an individual has been completely immobile and whether a predetermine critical threshold of concern ($L_F = 20$) is approached when an alarm would be raised.

(iv) Localized Movement

There are situations when an event may be localized but with small-scale irregular movements superimposed. For example, in Figure 11.2.1.3c, an individual may be so critically placed that a small arm movement is sufficient to be alternately detected/not detected at one location (n = 2) but not enough to activate an adjacent location (n = 1 or 3). Conversely, the movement may alternately activate each of two adjacent points (n = 2 and 3, Figure 11.2.1.3d). Fluttering of curtains due to convection, restless animate objects, vibration of mechanical devices, etc., are examples of such localized movements.

In order to accommodate such situations, it is necessary to define scales of movement in terms of the size of the region concerned, for it to be classified as a "small-region movement."

The distance moved by an event shifting from one location to another may be quantified in terms of the change in the value of the spatial chromatic parameter H_s (Section 11.2.1.3 (iii), Figure 11.2.1.4a). Thus, the extent of a region covered by a movement may be defined by a maximum H variation within a given time period.

The difference between two H values (H_r and H_i) can be quantified as the length of the chord d_H between these H values on an H-S polar diagram of unit radius of the kind introduced as Figure 1.4 in Chapter 1, modified to show d_H as Figure 11.2.1.5.

$$d_H = [2 - 2\cos(H_i - H_r)]^{1/2} \tag{5}$$

Figure 11.2.1.3b shows that each of the R, B processors covers $w + 1$ locations. On the H-S diagram (Figure 11.2.1.5b), the H range covered by each of the R, B processors is 120 (R + B range = G range = 240). A *small region* may be defined as n_s the number of consecutively sorted locations that corresponds to a $H_i - H_r$ range of $120(n_s - 1)/w$. The threshold for determining the extent of a small region d_{SR} in chromatic space is then given from Equation 11.2.1.10 as

$$d_{SR} = \left\{ 2 - 2\cos\left(\frac{120(n_s - 1)}{w} \right) \right\}^{1/2} \tag{6}$$

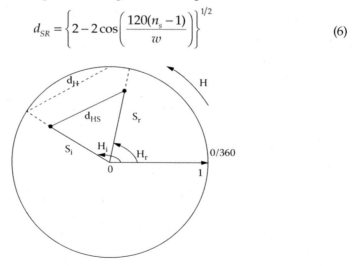

FIGURE 11.2.1.5
H-S polar diagram showing d_H as the chord on a unity radius circle. (From Xu S. and Jones G. R. (2006). Event and movement monitoring using chromatic methodologies, *IoP J. Meas. Sci. Tech.*, 17, 3204–3211. With permission)

(v) Deployment Examples

Some typical examples of the deployment of the technique without and with small-region considerations are shown, respectively, in Figures 11.2.1.4a and b, in which the horizontal axes represent time [Xu and Jones 2006].

For the case that ignores small-region movement (Figure 11.2.1.4a), two parameters are shown—the H value and corresponding location (left vertical axis), and L_F corresponding to residence time (right vertical axis). There are five distinct time regions—A: a region of continual activity, B: immobile period, C: short-duration highly localized period, D: immobile period, E: continual activity. During the immobile periods the clock increments the L_F values and resets once a movement is detected (e.g., C region) so that the duration of immobility does not reach a critical level indicated by the immobility threshold ($L_F = 20$).

In the case that takes account of small-region movements (Figure 11.2.1.4b), the two parameters shown are the size of the movement d_H (Equation 11.2.1.10) derived from the H values (left vertical axis) and the small-region movement duration (right vertical axis). The time increments ignore the resetting if only a small region movement is detected. Division of the test results into distinct time regions now only identifies three regions—A, B+C+D, and E—rather than the five identified in Figure 11.2.1.4a. The difference is that the small-region movement at C does not reset the clock, and so the immobility duration continues to increase and reaches a critical immobility threshold (60) at the end of the B+C+D period.

(vi) Summary of Processing Procedures and Capabilities

Implementations of the various chromatic processing procedures described in previous sections are summarized in the flowchart in Figure 11.2.1.6. Not only does this show the sequential application of the optical and spatial chromatic processing for movement monitoring but also indicates the need to distinguish between large-scale movement for indicating activity levels and small-scale movement associated with immobility. The diagram also emphasizes the capability, via the optical chromaticity, of monitoring specific parameters and events such as temperature (via thermochromic elements, Chapter 6, Section 6.3.4.2; Chapter 9, Section 9.3.2.2), television set status, door status, ambient light changes, etc.

These latter capabilities are not only significant for status evaluation but also for discriminating between inanimate and person-induced events. For example, small-scale events at a point located on a curtain is indicative of the curtain fluttering and not the immobility of a fallen individual. Various rules covering a number of such event activities may be incorporated into the decision-making software in accordance with the demands of different applications.

11.2.2 PIR Augmented Systems

The optically based chromatic system (Section 11.2.1) may be augmented through the use of PIR Sensing (Wong 2006). This gives the advantage of

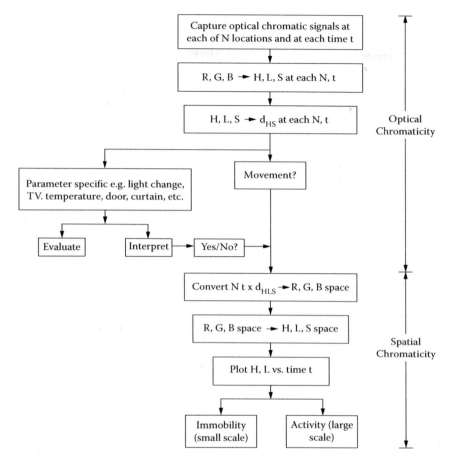

FIGURE 11.2.1.6
Chromatic processing in the optical and spatial domains. (From Xu S. and Jones G. R. (2006). Event and movement monitoring using chromatic methodologies, *IoP J. Meas. Sci. Tech.*, 17, 3204–3211. With permission.)

providing additional cross-correlation means for distinguishing between inanimate objects and individuals, for eliminating artifacts such as ambient light changes, and for providing preliminary indications under dark conditions before triggering the operation of the optical system.

PIR sensors detect changes in the infrared emission within an environment on relatively short time scales. They are therefore able to detect the movement of an individual through changes in the infrared radiation due to body heat emitted by the individual. Consequently, in their normal mode of operation, such sensors do not respond to stationary individuals nor other steady sources of infrared radiation nor to gradually changing levels of such radiation. They are essentially movement detectors; hence, their primary use in burglar alarms.

There are two forms of PIR-enhanced systems that can be deployed. The first utilizes a single PIR unit that can be operated normally in a continuous mode or that can be rapidly triggered via a shutter to produce a detectable change in infrared level from a stationary infrared limiting object.

The second form utilizes a trefoil arrangement of PIR sensors to provide two-dimensional spatial discrimination similar to that described for tag location (Chapter 2, Section 2.4) and for optically monitoring electric arc plasma columns (Chapter 4, Section 4.3). In this mode of operation, each of three PIR sensors have their spatial responses partially overlapped, as a result of which the spatial discrimination is increased from three sectors with three orthogonal sensors to seven sectors when the sensors are deployed nonorthogonally (the seven-sector limit results from there being no variability of response with angular position in the PIR case).

11.2.2.1 *Single-Triggered PIR Unit*

The single PIR unit operates with a triggered shutter that is normally opened but can be activated to close and then open to provide the transient conditions for the PIR to respond.

A typical train of signals produced by a triggered PIR unit is shown on Figure 11.2.2.1a, which distinguishes between the direct PIR operation and triggered operation with and without a received signal. Whereas the PIR system simply addresses the entry area indiscriminately in a digital manner, the output from the optical chromatic unit discriminates between different spatial locations, and the signal for each location is analogue in form. The latter may be converted to a digital indication of locations occupied at various times as shown on Figure 11.2.2.1b. The occupied locations are determined as described in this section from changes in the chromatic signatures of the visited locations, followed by the application of spatial chromatic processing. (Section 11.2.1.3). This shows periods of quasi-continuous movement (around 14:26:54, 14:29:02, 14:32:37) interspersed by periods of movement.

The optical chromatic results of Figure 11.2.2.1b are cross-correlated with those of the PIR unit (Figure 11.2.2.1.1a) in Figure 11.2.2.1c. This is achieved by showing digitally three parameters—whether an individual was known to be present (0/1), whether there was an optical chromatic change (1/2), and whether there was a PIR signal, direct or indirect (2/3).

A decision table (Table 11.2.2.1) is used in conjunction with the data on Figure PIR-1.1c for correlating the results. Thus, a direct PIR signal without a corresponding optical signature (14:24–16/24, A) indicates a transient heat source rather than the presence of an individual, whereas an optical signal without a PIR signal (14:24:40B) indicates a light change or inanimate object rather than an individual. During time periods of inactivity (e.g., 14:29:55–14:31:45, Figure 11.2.2.1b), the optical unit records a chromatic signature change invariant with time, and the direct PIR unit does not register a signal (Figure 11.2.2.1c), but when triggered repeatedly, does so. The PIR and optical outputs therefore correlate to confirm the presence of an immobile individual.

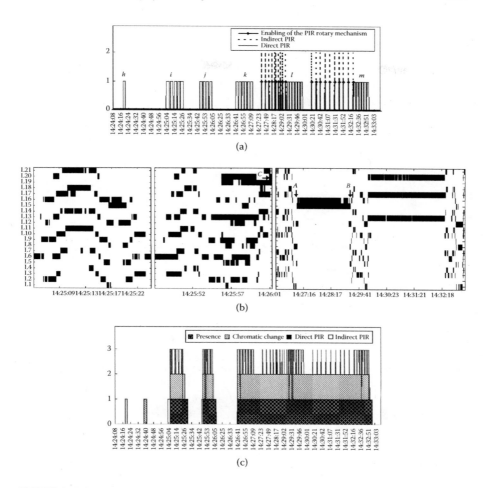

FIGURE 11.2.2.1
Optical chromaticity augmented by a single PIR: (a) binary signals for PIR sensor (h—temperature change; i,j,k,l,m—activities of individual), (b) locations at which chromatic changes occurred (0–21 locations), (c) cross-correlation of binary chromatic change, direct and indirect PIR response.

There are small but finite movements to which the direct PIR unit does not respond but both the optical and indirect PIR systems do. Such situations may be indicative of a fallen, but not immobile, individual or an individual seated on a chair, for example. Once the triggered PIR has confirmed the presence of an individual (and not, for example, a slightly fluttering curtain), the scale of the individual's activity may be determined from the range of locations showing chromatic signature changes. Figure 11.2.2.2 shows a record of the activity scale (normalized with respect to the recording points that are furthest apart) as a function of time. A threshold for small-scale activity is shown corresponding to 0.25—values less than this being classified as small scale.

TABLE 11.2.2.1

Decision Table for Cross-Correlating Results from Optical
Sensor Array and PIR Sensor Module for Presence Detection

Initial Presence State	Chromatic Change	PIR Response Direct	PIR Response Indirect	Presence Decision
N	1	1	—	Y
N	0	1	—	N
N	1	0	—	N
N	0	0	—	N
Y	1	1	—	Y
Y	1	0	1	Y
Y	1	0	0	N
Y	0	—	—	N

Note: "N": no presence; "Y": presence; "—": signal not needed.

When tracking whether an individual is mobile or not, if activity remains continuously below this level for a prolonged period and the location is on the floor, then an alarm may be raised after a predetermined period.

The performance of such a combined chromatic/single PIR system has been shown, from a limited number of tests, to be 100% reliable in detecting critical inactivity in a quiescent environment and 73% reliable when artifacts such as fluttering curtains, change of ambient light, etc., occur. Although such a performance is acceptable for many practical situations, improvements are achievable. For example, uncertainties with respect to the triggered PIR responding to a fixed heat source (e.g., radiator, etc.) rather than an immobile individual may be addressed through the use of a thermochromic element being attached to the source and being chromatically read by the optical system to provided temperature values (Chapter 6, Section 6.3.4.2; Chapter 9,

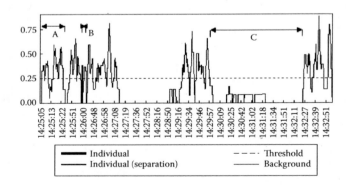

FIGURE 11.2.2.2
Relative activity as a function of time showing the threshold for small-scale activity.

Section 9.3.2.2). A further cross-correlation between the PIR and optical units should then alleviate the problem.

11.2.2.2 PIR Units Operated in Space Chromatic Mode

Chapter 2, Section 2.4 and Chapter 4, Section 4.3 show how three detectors deployed in a trefoil mode may be used for determining the location of a source in two-dimensional space. This requires that the angular response lobes of the detectors be nonorthogonal with respect to each other. Three PIR units may be deployed in this manner for providing a level of spatial correlation with an optical chromatic unit for addressing the activity of an individual. A schematic diagram showing such a deployment of three PIR units is shown in Figure 11.2.2.3a overlaid on an area to be addressed (Wong 2006). Unlike the cases described in Chapter 2, the responsivity of each PIR is not dependent on angular direction, so that chromatically this constitutes a special degenerate case.

Figure 11.2.2.1a shows that the nonorthogonality of the trefoil arrangement leads to the area addressed being divided into seven discrete zones (R, G, B, C, P, Y, W). Within each of the seven zones, there is no position discrimination because of each detector responsivity not varying with angular direction. Consequently, spatial discrimination is coarse and limited to each of the seven discrete zones. Thus, the effect of the nonorthogonal deployment is to increase the spatial discrimination from three zones with three orthogonal detectors to seven zones with three nonorthogonal detectors. This constitutes a more efficient use of three detectors to improve the spatial resolution by a factor of 2.3. Increasing the number of detectors from three to four (e.g., Chapter 2, Section 2.3) would increase the number of zones in orthogonal mode to four and in nonorthogonal mode to thirteen, i.e., a nonlinear increase in spatial resolution.

A further difference between the PIR nonorthogonality and those described in Chapter 2, Section 2.4, and Chapter 4, Section 4.3, is that the lobe geometries of the former are curved, whereas, with the PIR, the zone sectionalisation is rectangular (Figure 11.2.2.3b). This arises from the PIR-sensing element being rectangular and highly collimated.

Discrimination between the presence of an individual in each of the seven addressed zones of Figure 11.2.2.3b is achieved with the decision table, Table 11.2.2.2. This shows how signals from each detector (R, G, B) are interpreted in terms of spatial chromaticity (Figure 11.2.2.3) to yield H and L values. For example only a signal from the R PIR (1) and none from the G and B PIR (0) indicates a presence in zone R corresponding to $0 < H < 60$, $L = 0.3333$, whereas signals from both R and G PIR(1) but not B(0) indicates a presence in Zone Y with $60 < H < 120$, $L = 0.6667$, etc.

As an animate object moves across each of the seven regions shown in Figure 11.2.2.3b, the source can be tracked in terms of the H variation with time. An example of such movement tracking is shown in Figure 11.2.2.4. This provides a convenient means for not only determining the location of an

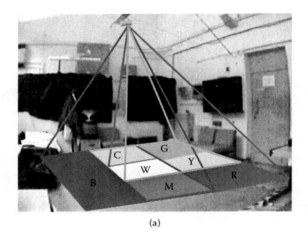

(a)

PIR 'B'   PIR 'G'

(b)

FIGURE 11.2.2.3
(See color insert following page 18). Deployment of PIR in trefoil space chromatic domain: (a)
isometric view. (b) plan view.

TABLE 11.2.2.2

Conversion of Spatial PIR R, G, B to *H, L, S*

PIR R	PIR G	PIR B	Hue, H	Lightness	Zone
0	0	0	0	0	None
1	0	0	0	0.3333	Red
1	1	0	60	0.6667	Yellow
0	1	0	120	0.3333	Green
0	1	1	180	0.6667	Cyan
0	0	1	240	0.3333	Blue
1	0	1	300	0.6667	Magenta
1	1	1	360	1	White

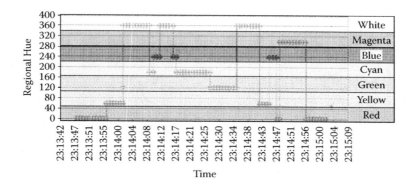

FIGURE 11.2.2.4
PIR hue-based location as a function of time showing the movement of an individual between the seven PIR zones.

individual but also the pattern of movement across the seven regions being addressed.

The spatially resolved PIR signals can be cross-correlated in greater detail than the single-PIR unit (Section 11.2.2.1) with the spatially resolved optical chromatic signals. Thus, the decision table (Table 11.2.2.2) may be applied to each of the seven PIR Chromatic regions of Figure 11.2.2.3. Additionally, there is the possibility of allocating the spatial regions of the PIRCS and optical chromatic units to regions with specific functionalities. For example, PIR zones may be allocated so as to avoid heat sources that might produce ambiguous signals; both PIR and optical zones may be organized to address in conjunction entry and door regions so that responses from both units together confirm that not only has a door been opened but also that an individual has entered/exited. The decision table may therefore be extended to include a number of such cross-correlation possibilities between the PIR and optical signals.

Tests with such a system have yielded an overall success rate of 94% in identifying events correctly [Wong 2006].

The information compression achievable through the use of chromatic methods in the space domain enables a considerable range of events to be economically addressed. It leads to the possibility of addressing several sites via multiplexing between various PIRCS/optical units.

Site tests have shown the capability of multiplexing between at least seven rooms simultaneously in real time using PIRCS-correlated optical units in chromatic space modes [Wong 2006, Smith 2003, Smith 2005]. The volume of information processed is reflected by the fact that nine million parameter values are addressed each day by the system that operates continuously over several months. Levels of activity are expressed in terms of space chromatic lightness. Figure 11.2.2.5 [Yang 2007] shows a typical record of weekly activity for a single individual over a period of 11 months. This allows trends in inactivity variations to be mapped either hourly or weekly for

TABLE 11.2.2.3

Examples of Daily Activity Levels Derived from Figures of the Form of Figure 11.2.2.5 for the Occupants of 10 Rooms

Date	RM1	RM6	RM9	RM12	RM18	RM19	RM28	RM33	RM34	RM38
09-16	Agile	/	Normal	Normal	Agile	/	Normal	Normal	Normal	Normal
09-17	Normal	/	Normal	Agile	Normal	Normal	Normal	Immobile	Normal	Normal
09-18	Normal	/	Normal	Generally agitated	Normal	Locally agitated	Less Mobile	Agile	Agile	Less Mobile
09-19	Normal	/	Less Mobile	Normal	Normal	Agile	Normal	Less Mobile	Agile	Normal
09-20	Less Mobile	/	Normal	Agile	Agile	Normal	Normal	Normal	Normal	Normal
09-21	Normal	Normal	Normal	Agile	Normal	Normal	General agitated	Agile	Agile	Agile
09-22	Less Mobile	/	Normal	Immobile	Less Mobile	/	Normal	Agile	Less Mobile	Normal

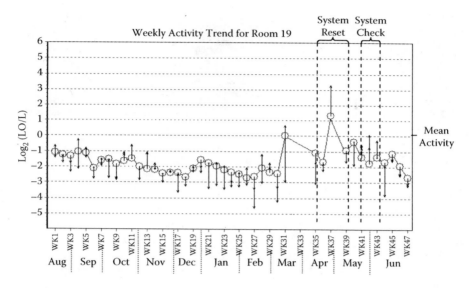

FIGURE 11.2.2.5
Weekly activity trend for a single room (August 2005–June 2006). (Arrows → dominant location, spread of locations). (Figure courtesy C Yang [2007]).

prolonged periods. The arrows are measure of H (indicating the dominant location) and S (showing the spread of locations visited).

Table 11.2.2.3 shows an example of tabulation of the level of activity derived from graphs of the form given in Figure 11.2.2.5 for the occupants of 10 rooms for each of the event days. This allows deviations from normal activity for each individual to be estimated.

11.3 Physiological and Physical Indicators of Fatigue

11.3.1 Introduction

There is an increasing awareness of the influence of fatigue in reducing the effectiveness of an individual in undertaking tasks such as vehicle driving. Drowsy driving encompasses several aspects, which include falling asleep or lacking concentration, both of which can lead to fatal accidents [Reyner and Horne 1998, Horne and Reyner 1995a] (Figure 11.3.1.1).

Methods for assisting drivers to recognize indicators of the onset of disruptive fatigue are essential for preventing fatigue-related road accidents, which claim 1 in 5 of UK motorway accidents [Horne 2001].

FIGURE 11.3.1.1
Hazard of driving when fatigued.

One such system is the Advisory System for Tired Drivers (ASTiD). This system incorporates the superposition of a circadian rhythm [Horne and Reyner 1995b] upon signals from a gyroscope arranged to respond to the lateral movement of a vehicle [Koh et al. 2007]. The circadian rhythm is a diurnal variation in the level of alertness of a human. Amongst the lateral movements detected by the gyroscope are those produced as swerves in response to the impulsive reactions of a driver emerging from periods of micro-sleep. Also, there are preceding quiescent periods consequent upon a fatiguing driver taking fewer minor corrective steering actions.

The output signal from an ASTiD unit is in the form of a time-varying voltage composed of the addition of three components: the circadian rhythm curve, a component whose magnitude increases with duration of continuous driving, and the gyroscope output (Figure 11.3.1.2). This particular

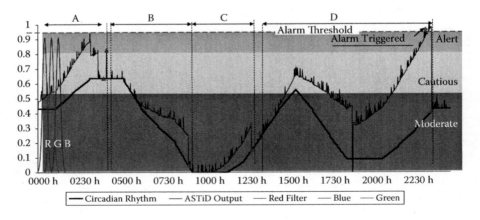

FIGURE 11.3.1.2
Typical output of the ASTiD system from a typical 24-h period test with several changes of drivers. (A, B, C, D different driver and shift periods.) (From Koh A., Jones G. R., Spencer J. W., and Thomas I. (2007). Chromatic analysis of signals from a driver fatigue monitoring unit, *Meas. Sci. Technol.* 18, 1–8. (With permission.)

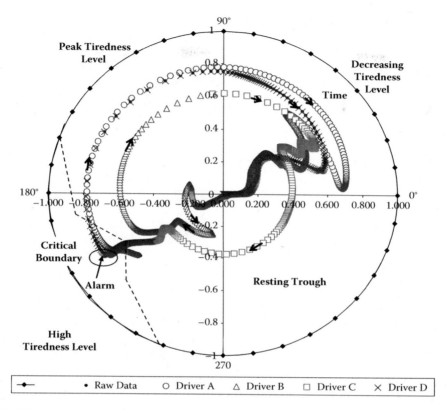

FIGURE 11.3.2.1
(See color insert following page 18). Chromatic polar diagram H-L for the overall output signal of Figure 11.3.1.2. (Different symbols corresponds to various drivers A–D, Figure 11.3.1.2) (From Koh A., Jones G. R., Spencer J. W., and Thomas I. (2007). Chromatic analysis of signals from a driver fatigue monitoring unit, *Meas. Sci. Technol.* 18, 1–8. With permission.)

record covers a period of 24 h during which the vehicle was driven by three drivers and a high fatigue level being ultimately reached at 22.33 h. The benefits provided by such a system can be enhanced by improving the information extraction and correlation between the physiological and physical indicators.

11.3.2 Physiological Indicators

The time variation of the circadian rhythm is shown on Figure 11.3.1.2 as a periodic continuous curve, whereas the contribution due to continuous driving is added to this curve to give the overall curve (neglecting the higher frequency pulses). Consequently, this overall curve represents the total physiological signal.

The total physiological signal may be chromatically transformed by time stepping (1-min step) three chromatic processors (75-min wide) along

the signal (e.g., Chapter 10, Section 10.2, Figure 10.2.1 (iv)). The resulting chromatically transformed signal is shown on the H-L diagram of
Figure 11.3.2.1 [Koh et al. 2007] where the different symbols represent each
of the four stages (A–D; Figure 11.3.2.1). There are a number of phases to
the curves that can be associated with different aspects of the physiological
curves. Within the second quadrant (90–180°) there are three arcs (A, B, D)
that represent a change from increasing to decreasing fatigue. Within the
fourth quadrant (270–360°) there is a single arc (c) that represents the rest-
ing trough.

The approach toward excessive fatigue is indicated by values of the H-L
coordinates 210°:0.8, which can raise an alarm [Koh et al. 2007]. A threshold
curve can be established mainly within the third quadrant (180 < H < 270°)
and with H ~ 0.8, which serves to define the physiologically determined fatigue
boundary.

11.3.3 Physical Indicators

The output signal from the gyroscope is in the form of a series of pulses
as shown in Figure 11.3.1.2 superimposed upon the overall physiological
curves. An expanded example of the gyroscope output in isolation and
over a smaller time period shows the existence of a number of different
pulses produced by various road conditions (Figure 11.3.3.1a). These include
signals when negotiating speed bumps, roundabouts, etc., as well as fatigue-
simulated swerves.

Signals of the form shown in Figure 11.3.3.1b [Koh et al. 2007] may be chromatically transformed by time stepping three nonorthogonal processors
(R, G, B; Chapter 10, Section 10.2, Figure 10.2.1 (iv)) along the time axis. Such a
procedure shows that the fatigue-related swerves have particular chromatic
H and S parameter variations that distinguish them from other gyroscope output signals. This results from the particular manner in which the
vehicle steering wheel is moved during the impulsive response of a driver
emerging from a micro-sleep period. Figures 11.3.3.2 a and b show the correlation between the time variation of the steering-wheel acceleration and
velocity with chromatic saturation and lightness, respectively.

For real-time, online operation, it is sufficient to set a threshold for the
magnitudes of chromatic L and S above which the signal is regarded as
being due to fatigue-induced swerving. An example of the application of
this condition to real-time gyroscope pluses, transformed to S and L as
functions of time, shows that fatigue-induced signals can be identified
(Figure 11.3.3.3).

A number of tests have shown that the L and S thresholds apply regardless of
change of driver, vehicle, weather conditions, or traffic density [Koh et al. 2007].

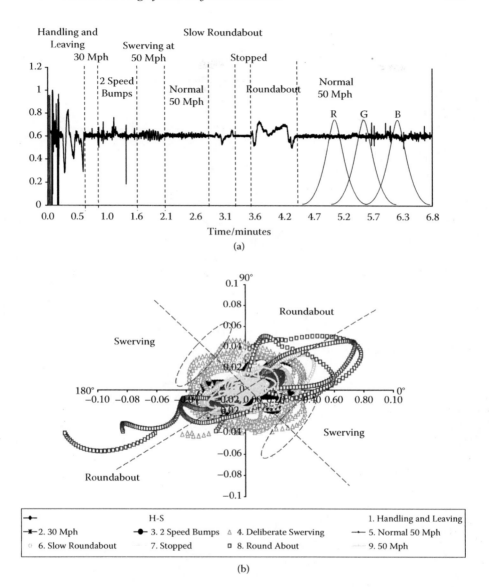

FIGURE 11.3.3.1
(See color insert following page 18). Chromatic processing of gyroscope signals: (a) gyroscope output showing different signals associated with various driving features. (b) H-S polar diagram of the gyroscope output of Figure 11.3.3.1(a) (different symbols indicate different maneuvers). (From Koh A., Jones G. R., Spencer J. W., and Thomas I. [2007]. Chromatic analysis of signals from a driver fatigue monitoring unit, *Meas. Sci. Technol.* 18, 1–8. With permission.)

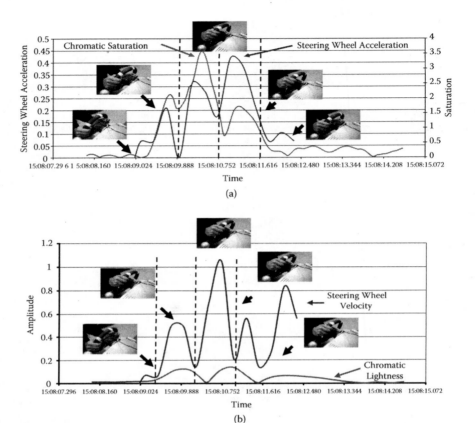

FIGURE 11.3.3.2
Correlation of gyroscope signals with steering-wheel movement: (a) correlation between
the time variations of steering-wheel acceleration and chromatic saturation, (b) correlation
between the time variations of velocity and lightness.

11.3.4 Correlation between Physical and Physiological Indicators

Whereas the physiological indications of fatigue are that the relevant L and
H chromatic parameters should exceed their thresholds value, the physical
indications on L and S not only need to exceed their threshold but need to
do so a number of times. Correlation of the physiological and physical indicators may be considered in terms of chromaticity-based probability factors
[Zhang et al. 2005], whereby the physiological factor depends upon L and H,
but the physical factor is based upon the number of L and S identified pulses
in a given time period.

This illustrates how a chromatic approach improves upon the simple expe-
dient of summing the amplitudes of signal voltages without regard to distinguishing the cause of the latter (Figure 11.3.3.3).

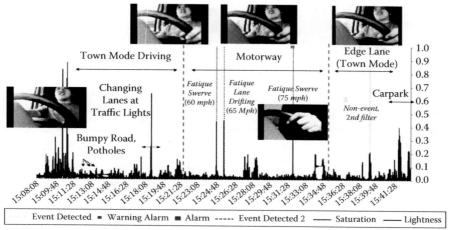

Number of Events Detected in 35 minute duration is 3

FIGURE 11.3.3.3
(See color insert following page 18). Real-time chromatic analysis of the gyroscope output.

11.4 Summary

The chromatic methodology may be deployed for addressing behavioral trends of an individual with minimal intrusion on personal privacy but in an economical and information-efficient manner. Two examples (assistive care of the elderly and infirm, fatigue of road vehicle drivers) illustrate some of the different ways in which the methodology can be applied.

The assistive care application illustrates

- The use of chromatic signatures, in the optical domain, of a limited number of locations (~20) for indicating the presence of objects/persons (from the points eclipsed) and addressing specific functions at particular locations (e.g., temperature, TV on/off, door opening, etc.)

- The use of the chromatic space domain in conjunction with the time change of optical domain chromatic signatures at various locations for qualifying the general activity of an individual in terms of L (level of activity), H (dominant location of activity), and S (the extent to which activity is distributed amongst different locations)

- The combination of optical and space-domain chromaticity with PIR space chromaticity to enhance the distinction between the movement of an individual and artifacts (such as ambient light changes, fluttering curtains, etc.), the level of immobility of an individual, the control of light switching for different levels of activity checking at nighttime, etc.

- The conservative approach to information and system economics that enables multiplexing of several sites to occur within acceptable time scales and costs.

These various aspects are summarized in Table 11.4.1
  The driver-fatigue application indicates that

- The use of time-stepped chromatic processing of the lateral movement of a vehicle can distinguish physiologically induced responses from drivers becoming increasingly fatigued independent of person, vehicle, or weather.

TABLE 11.4.1

Comparison of Individual and Combined Capabilities of Various Chromatic Domains

Chromatic Domain	Individual Capabilities	Combined Capabilities
Optical (wavelength)	• Detailed locations • Specific functions (e.g. Door, TV, Temperature, etc.) • Moderate time response	• Improved immobility detection • Some night operation improvement (inc. light control) • Improved heat discrimination / • Good time/location compromise • Activity/immobility confirmation • Much improved night operation (inc. light control) • Improved artifact discrimination
Optical (Spatial)	• Detailed activity • Immobility (inc. fall) • First level artifacts (e.g., ambient light, Inanimate objects)	
Single PIR (none)	• Moving person (NO spatial resolution) • Stationary person • Rapid time response	
Trefoil PIR (spatial)	• Moving person • Coarse location/activity • Convenient night operation • Rapid time response	

- Empirically determined fatigue contributing factors (e.g., circadian rhythm, period driving) may be transformed with chromatic time-domain stepping to provide indications of fatigue variations.
- The trend towards high levels of fatigue may express in terms of probability functions based upon chromatic parameters form the vehicle movement and empirical considerations.

A summary of these aspects is given on Table 11.4.2.

TABLE 11.4.2

Chromatic Parameters of Physical and Physiological Indicators of Fatigue

Chromatic Domain	Fatigue Indicator	Chromatic Parameters	
		Individual	Combined
Time stepped	Physical (lateral movement)	Number of swerves (N) with specific L, S values above threshold	
Time stepped	Physiological (circadian rhythm, driving period)	H_{CR}, S_{CR} approaching threshold levels	

References

Carron T. and Lambert P. (1994). Color edge detector using jointly hue, saturation and intensity, *IEEE Int. Conf. on Image Processing* (Austin, TX), pp. 977–981.

Dony R. D. and Wesolkowski S. (1999). Edge detection on color images using RGB vector angle, *Proc. IEEE CCECE'99* (Edmonton, May 9–12) Canada.

Horne J. A. (2001). Sleep-related vehicle accidents, *Some Guidelines for Road Safety Policies. Transportation Research* (Elsevier) — Part F, 4: 63–74.

Horne J. A. and Reyner L. A. (1995a). Driver sleepiness, *J. Sleep Res.*, 4 (Suppl. 2): 23–29.

Horne J. A. and Reyner L. A. (1995b). Sleep related vehicle accidents, *Br. Med. J.*, 1995, 310, (6979): 565–567.

Koh A., Jones G. R., Spencer J. W., and Thomas I. (2007). Chromatic analysis of signals from a driver fatigue monitoring unit, *Meas. Sci. Technol.* 18, 1–8.

Reyner L. A. and Horne J. A. (1998). Falling asleep whilst driving: Are drivers aware of prior sleepiness?, *Int. J. Legal Med.*, 111: 120–123.

Smith D. H. (2003). Video Multiplexer Unit, *Private Communication*.

Smith D. H (2005). Mechanical Fixtures for PIRCS, *Private Communication*.

Wesolkowski S. and Jernigan E. (1999). Color edge detection in RGB using jointly Euclidean distance and vector angle, *Proc. IAPR Vision Interface*, Canada, pp. 9–16.

Wong K. (2006). Chromatic Monitoring of Living Environment, Ph.D. thesis.

Wong K., Xu S., and Jones G. R. (2005). Chromatic identification of complex movement patterns, *Proc. Int. Conf. on Complexity in Sci., Med. Sociol.*, Centre for Complexity Research, Liverpool.

Xu S. (2004). Chromatic System for Care Society, Ph.D. thesis, University of Liverpool.

Xu S. and Jones G. R. (2006). Event and movement monitoring using chromatic methodologies, *IoP J. Meas. Sci. Tech.*, 17: 3204–3211.

Yang C. (2007). Occupant's Behavioural Pattern Measurement, Private Communication.

Zhang J., Jones G. R., Deakin A. G., and Spencer J. W. (2005). Chromatic processing of DGA data produced by partial discharges for the prognosis of HV transformer behaviour, *Meas. Sci. Technol.*, 16: 556–561.

Epilogue

An illustrated overview of some of the key uses of the generic methodology of chromaticity for monitoring complex systems has been provided. Some of the main functions and capabilities that the chromatic approach can provide may be summarized as follows.

Supervision—Different levels of organization (or different domains) of system behavior may be consolidated using the same technology (e.g., polar plots) in an embedded sense, providing an overall, holistic information "picture." This is analogous to, and may be converted into, top-level information, such as a red/amber/green status for overall system behavior, which is abstracted from the complex, fuzzy world to enable decision making. There are parallels here with, for instance, an operator's or pilot's view. Simon refers to decision making on the basis of incomplete or uncertain information as "satisficing" (Simon 1957), which is decision making based on using available information, whether it is complete or not, to reach a decision. Chromatic monitoring goes beyond this, having extracted the essential information from the complex web of data that is available, while retaining the traceability of significant information extracted at source in case it is needed for subsidiary views. By analogy, the pilot focuses on the main top-level indicators to assess overall system health such as speed, direction, altitude, attitude, location, and fuel availability, but has access to further, specific, more minor instruments and data that expand on or support the higher levels if needed.

View—Chromatic monitoring has the capability for zooming in to micro scales for fine detail (e.g., local weather) or out to the whole system behavior (e.g., regional, global weather patterns, global timescales) as required, accomplished in part through filter design.

Tracking—Monitoring system behavior against plan or expectation or for abnormal behavior, which may be applied to theoretical or empirical (or hybrid) system models. This is an extension of supervision in that ideal or expected system behavior is already known and may be incorporated into the monitoring, e.g., as a behavioral trajectory on the polar plot. It is then possible to predict when intervention in the system is required.

Statistics—With filters based on Gaussian characteristics and the Gabor transform basis, chromatic analysis effectively compares the data to statistical distributions (e.g., the normal distribution), with H equivalent to the mean, S equivalent to the variance, and L equivalent to amplitude (energy). In particular, rare but potentially catastrophic events that may occur in the "tails" of the normal distribution are amplified by means of the overlapping filters, which can perform an initial cross-correlation by means of normal distributions overlapping at the tails. Also, it has been shown that virtually all signals may be represented with only 3–6 parameters, whereby the increase in discrimination from 3 (e.g., R,G,B) to 6 is minimal in practical terms (Jones et al. 2000). Ongoing work will also show that chromatic processing may yield an approximation of the information content of a data source (its entropy).

Multiple domains—For example time, frequency, and spatial distribution are suitable for concurrent chromatic processing and can aid cross-correlation of system behavior, if required. It may be shown that this approach contributes more than alternative techniques such as the more limited (single-domain) cases of Fourier analysis and wavelet analysis.

Multiple sources—Chromatic processing may be applied across the spectrum, e.g., directly to data from acoustic/visual/optical-fiber sensors, which again may assist cross-correlation of the holistic picture of system behavior, or even directly to derived sources in the information domain, e.g., to financial information.

Transparency—Compared with some AI techniques such as neural nets, for example, chromatic processing directly supports traceability of the characterization of system behavior.

In summary, chromatic analysis is useful as an interdisciplinary bridge or metaphor for investigating diverse complex systems and consolidating the range of information about system behavior that is available, from direct physical parameters to holistic system overviews. Also, its polar plots may be useful as an intermediate human interface to complex system behavior. Chromatic processing can produce a qualitative characterization of a system (behavioral trajectory) and additionally *quantify* this using, for example, secondary HLS analysis and deriving event probability measures (Zhang et al. 2005).

References

Jones, G. R., Russell, P. C., Vourdas, A., Cosgrave, J., Stergioulas, L., and Haber, R. (2000), The Gabor transform basis of chromatic monitoring, *Meas. Sci. Technol.,* 11, 489–498.

Simon, H. (1957), *Models of Man: Social and Rational,* John Wiley, New York.

Zhang, J. H., Jones, G. R., Deakin, A. G., and Spencer, J. W. (2005), Chromatic processing of DGA data produced by partial discharges for the prognosis of HV transformer behavior, *Meas. Sci. Technol.,* 16, 556–561.

Index

A

Acidity, 210–211,
Acoustical and vibration signals
  controlled-pulse train, 229–230, *229–230*
  diesel engines, 227, *228*
  double pulse signals, 230–232, *231*
  frequency domain, 224, 226–227, 229–236
  frequency-domain filters, 217–218,
    218–219, 220
  fundamentals, 213–214, 236
  hardware filters, 217–218, *218–219,* 220
  industrial plant faults, *225–226, 226–227*
  narrow-frequency-band continuous
    signals, 233
  processing deployment examples,
    224–236
  processing procedure selection, 214,
    215–216, 216–220
  pulsatile signals, *220–223,* 220–224
  random pulsatile events, 233, 235–236
  simulated pulse trains, 224, *225,* 226
  single pulse infrastructure, 232–233
  software filters, 217–218, *218–219,* 220
  time domain, 227, 233–236
Activities and behaviors, monitoring
  changes, selected locations, 243–248
  correlations, 262, *263*
  deployment examples, 248
  fatigue indicators, 257–262
  fundamentals, 239–240, *263–264, 264–265*
  monitoring system, 241–243, *242*
  movement, 245–247
  optical wavelength and space
    chromaticity, 240–248
  physical indicators, *258,* 260, *261–262*
  physiological indicators, *258–259,*
    259–260
  PIR augmented systems, 248–257
  single-triggered PIR unit, 250–253,
    251–252
  space chromatic mode operation, 253,
    254–257, 255, 257
  spatial chromaticity, *242,* 243, *244,* 245
  total chromatic change, *242,* 243
  vulnerable individual, 240–257

Advisory System for Tired Drivers
  (ASTiD), 258
Airborne microparticulates, 40
Airborne particles
  data complexity and processing, 40–42
  fundamentals, 192, 210
  particle types, *196,* 196–195
  portable systems, 192–193, *193*
  remote CCTV camera-addressed unit,
    194, *194–195,* 196
Aircraft
  fuel discrimination, 96–98, *97–98*
  inclination indications, 107–109,
    108–110, 111
  Lab transformation, 58
Air quality monitoring
  airborne particles pollution, 192–196
  air-quality data, 196–199, *198*
  fundamentals, 192
  particle types, 195–196, *196*
  portable systems, 192–193, *193*
  remote CCTV camera-addressed unit,
    194, *194–195,* 196
Algorithms, processing
  basic algorithms, 49–51, *50*
  denominator and numerator, gains, 56, *56*
  denominator term, gains, *50,* 55–56
  distimulus monitoring, *50,* 51–52
  fundamentals, 47–48, 58–59
  gains, *50,* 55–56, *56*
  generalized algorithms, 51–56
  H, S, V transformation, 48, *49*
  Lab transformation, *57,* 57–58
  tristimulus diagrams and monitoring,
    52–55
  tuned distimulus monitoring, 52, *53*
  x, y, z transformation, 49–56
Altitude changes, aircraft, 108
Anaerobic waste treatment
  cycle progression, *206,* 206–207
  fundamentals, 199–200, *200–202,* 211
  optical fiber sensing, 201, *201,* 203, *204*
  outputs, 200
  remote monitoring, *201,* 203, *205,* 206
  sensing and monitoring techniques,
    201–207

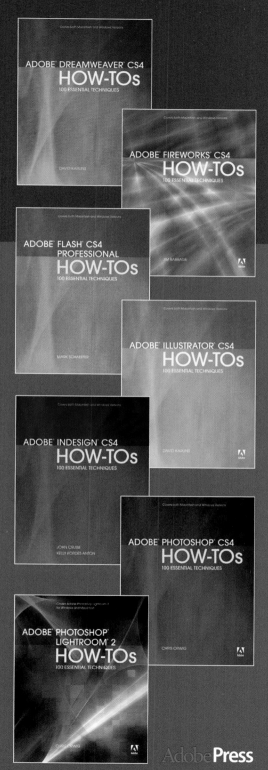